Dave's Electromagnetic Concept
(DEC)

David G Delgado

1

Table of Contents

3

Abstract

Dave developed a scalable Electromagnetic Concept (DEC) that can be used to several configurations and circumstances. The DEC's fundamental principle involves the spinning of metallic components to produce magnetic fields that act as forces on moving charges.

Wheelbarrows, drones, home generators, electric cars with no batteries, farm equipment, manufacturing, aviation, railroads, electronics, and auxiliary power units are just a few examples of the many objects that have applications. Paving the way for a huge number of prospective patents.

This manuscript is intended to assist you in comprehending the DEC ideas and how the system is expected to operate. In order to help readers understand the DEC concept using equations, Dave has included numerous examples. Some topics may be familiar to you while others may not. It may come in handy to have a basic understanding of calculus and unit vectors (Dot and Cross products).

Description of the "DEC Device"

Dave's concept consists of three basic parts:

1. Device consisting of a platform that revolves clockwise in conjunction with a central rechargeable battery starter pack in which the six capacitor plates are evenly placed and interspersed with the six regenerating accumulators that rotate in counter rotation.

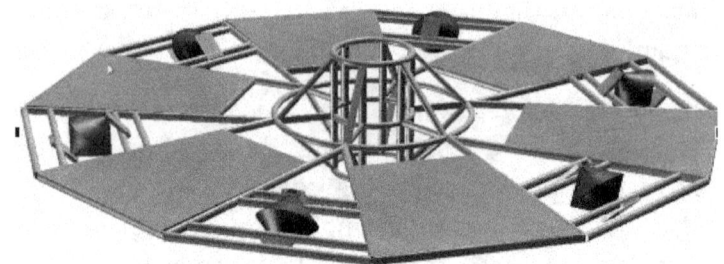

Figure 1 Internal Platform showing six capacitor plates evenly placed and interspersed with the six regenerating accumulators that rotate in counter rotation

2. Device consisting of a platform attached to the outside hull of the disc that rotates counterclockwise, as well as twelve electromagnets evenly placed and dispersed around the perimeter.

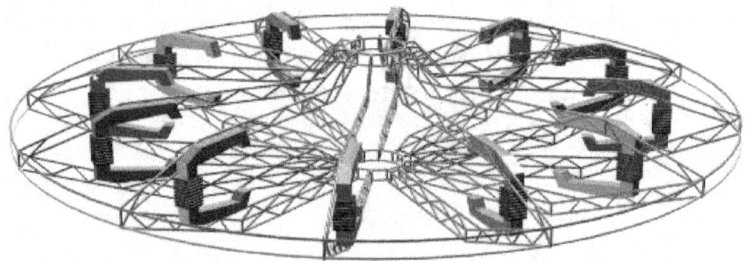

Figure 2 External Platform showing twelve electromagnets evenly placed and dispersed around the perimeter.

3. Stationary device which is anchored to the concept's center axis and to which the bottom support in the form of a tripod is attached and which, via bearings, forms a single set with the first two sections. Note, the spherical components shown are compressed air tanks used to initiate motion for the accumulators through nozzles. Standard air compressor included in stationary unit.

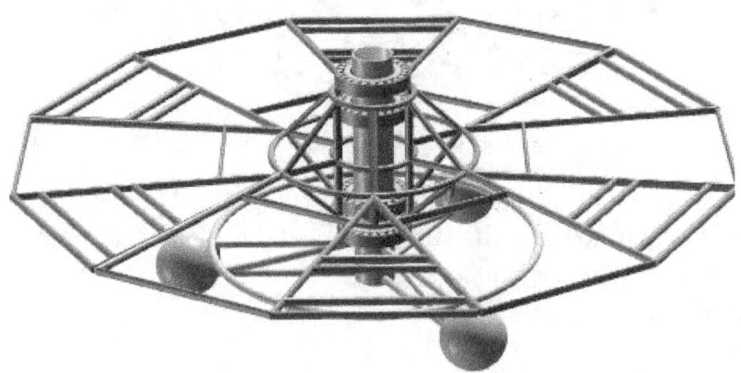

Figure 3 Stationary Platform

7

For reference, energy balance for nozzles and diffusers is written as follows:

$$m_{dot} (h_{in} + 1/2\ V_{in}^2) = m_{dot} (h_{out} + 1/2\ V_{out}^2)$$

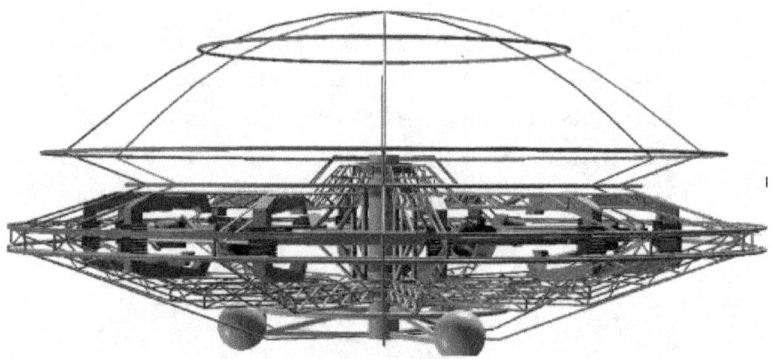

Figure 4 DEC side view

The illustration's framework is made out of a welded aluminum truss system to meet the strength and dimension requirements for a particular size and load as a point of reference from numerous structural choices.

The DEC animated computer graphic models as shown was created using Python Glowscript in vector format. Complete python code is included in an Amazon KDP 6x9 format book.

```python
# Inner Platform
Bi=sphere(pos=vec(0.26,0.46,0),radius=.01)
Ji=sphere(pos=vec(.65,1.13,0),radius=.01)
rp1 = extrusion(path=[Bi.pos, Ji.pos],shape=
shapes.circle(radius=.05))
rp2 = extrusion(path=[Ji.pos, Qi.pos],shape=
shapes.circle(radius=.05))
# Bicone  SCR
c=cone(pos=vector(0,0,0),axis=vector(.35,0,0),radius=.35,
color=color.red,opacity=.8)
run = True
def runner(r):
    global run
    run = r.checked
 checkbox(bind=runner, text='Run', checked=True)
scene.waitfor('textures')
t = 0
dt = 0.01
dtheta = 0.01
while True:
    rate(10000)
```

```
if run:
    p1.rotate(angle=-dtheta, axis=vec(0,0,1))
```

How Dave's DEC Works

DEC employs an internal platform that rotates in direct rotation, and multiple devices that rotate in counter-rotation with their axes fixed by bearings on this first platform. Six counter-rotating devices and regenerative accumulators are evenly positioned along the perimeter of the internal platform.

When the device works as a regenerative generator, another external platform rotates in counter-rotation to the first platform, containing twelve electromagnets that produce electrical energy during the passing of the accumulators in their air gaps.

This second external platform rotates in counter-rotation to compensate for the first internal platform's rotation and maintain the central cabin stationary. The inside platform functions as a rotor, and the external platform functions as a stator (armature).

One of the functions of accumulators and electromagnets is to make the system operate like a motor, with the rotor and stator (armature) freely rotating in opposite directions, keeping the cabin motionless in the center. The electromagnets are switched in a manner similar to current brushed motors, and this "motor" is powered by a central battery pack.

The speed of the motor can be controlled by varying the width and frequency of the applied pulses. There are a number of simple circuits used to provide the desired motor control pulses. One circuit includes a UJT/SCR control circuit, a UJT relaxation oscillator generates a series of pulses that drive an SCR on and off to vary the speed of the motor the UJT's oscillatory frequency is adjusted by changing the RC time constant. Another circuit may include a 555 timer that is used to generate pulses that drive a power MOSFET. An UJT/SCR control circuit schematic is shown as follows;

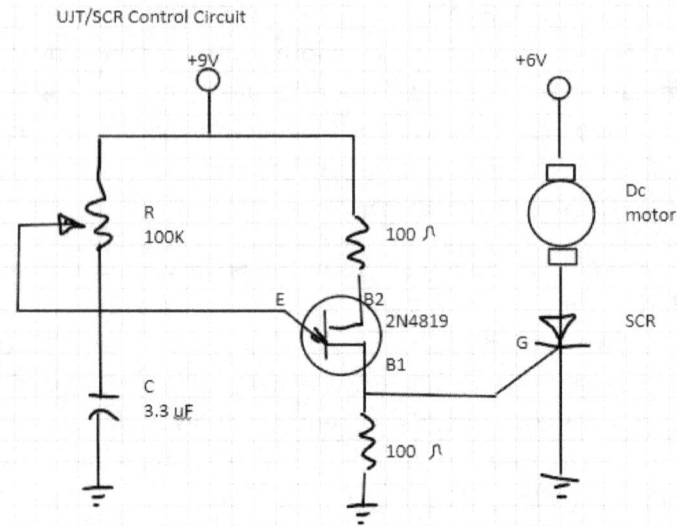

Figure 5 UJT/SCR control circuit schematic

When the system reaches a minimal rotation that allows the rotor (where the accumulators are) and the stator (where the electromagnets are) to work as a generator, switching the electromagnets allows energy regeneration of the batteries.

General relativity states that, in a specific limiting scenario, equations of the same form as in classical electromagnetism may be used to explain the gravitational field created by a rotating object (or any revolving mass-energy). The so-called "GEM equations" are the gravitational equivalents to Maxwell's equations for electromagnetic. They can be derived from the fundamental equation of general relativity, the Einstein field equation, under the assumptions of a weak gravitational field or relatively flat spacetime. This subject will be discussed in the section Gravitoelectromagnetism equations below.

In accordance with GEM equations, High-rotational bodies produce a gravitational field parallel to the spin axis and in the direction of the angular velocity vector. When two electrically charged bodies rotate in opposite directions on the same axis, the result is the formation of a gravitational center, which allows an object to be free of external gravitational forces

In addition, it is feasible for an object in high rotation to overcome the gravitational field force of the planet, and we performed calculations that allow us to quantify this phenomenon, known as the gyroscopic effect.

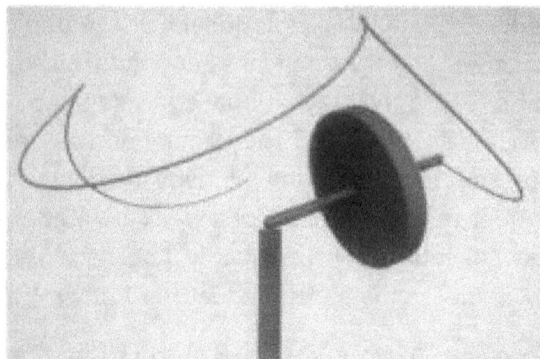

Figure 6 Illustration showing gyroscope with rotation outline

To resist the gravitational attraction of planet Earth, the platform's relative rotation speed would be equal to the Earth's rotation speed at the equator.

The platform's relative rotation speed would be equal to the Earth's rotation speed at the equator in order to withstand the gravitational attraction of planet Earth.

$U := 24901.55 \text{ mi}$

$U = 4.008 \cdot 10^7 \text{ } \cdot m$

$\omega := \dfrac{1}{24 \text{ hr}} \cdot \dfrac{\text{hr}}{60 \text{ min}} \cdot \dfrac{\text{min}}{60 \text{ s}} \cdot 2\pi$

$\omega = 7.272 \cdot 10^{-5} \text{ } \cdot s^{-1}$

$r := \dfrac{U}{2\pi}$

$r = 3.963 \cdot 10^3 \text{ } \cdot mi$

$r = 6.378 \cdot 10^6 \text{ } \cdot m$

$v := \omega \cdot r$

$v = 1.038 \cdot 10^3 \text{ } \cdot mph$

$v = 1.67 \cdot 10^3 \text{ } \cdot \dfrac{km}{hr}$

$a := \dfrac{v^2}{r}$

$a = 0.111 \cdot \dfrac{ft}{s^2}$

$a = 0.034 \cdot m \cdot s^{-2}$

with:

a = acceleration [m s⁻²]
υ = linear velocity [m s⁻¹]
ω = angular velocity [rad s⁻¹]
r = radius [m]
U = circumference [m]

Finally, the exterior magnetic system, which serves as a motor and generator, the magnetic system of the capacitor plates, which gives magnetic levitation, and the gyroscopic system, which provides gravitational propulsion, are included for effect as necessary.

The Electrical System

The Electric System includes a set of batteries, mounted inside the device initially feeds the entire electrical system of the craft consists of the electrical connection of countless electrical cells of any known type (lead-acid, nickel-cadmium, metal hydride, etc.). Standard electrical circuits and microprocessors are used for speed and directional control of DC motors. Again, when the system reaches a minimal rotation that allows the rotor and the stator to work as a generator, switching the electromagnets allows energy regeneration (charging) of the batteries.

Maxwell's Equations and Mathematical Models

Maxwell's Equations and Mathematical Models expose some principles involved in the rotation of metallic masses that allow to establish the conditions that satisfy the production of energy and propulsion:

1. Gravitational field induction in the direction of the angular velocity vector, that is, its axis of rotation, defined as Neutralization of Gravitational Potential in Gravitoelectromagnetism Propulsion Systems. This subject will be discussed in the section Gravitoinercial Propulsion System below.

2. Separation of electric charges from the electrolyte when the centrifugal potential energy exceeds the energy of the first ionization potential of the atoms that

16

make up the material, defined as Separation of Electrical Charges Using Gravitoelectromagnetism Propulsion Systems' Mass Rotation.

3. Separation of electric charges from the electrolyte when the rotating mass is subjected to a magnetic field, as a result of the Lorentz force $\vec{F}=q_E(\vec{v}\times\vec{B})$.

4. Gathering and projection of electrostatic charges from the atmosphere that flow to the magnetic vortex produced by the rotation of the magnetic field of the accumulator, defined as Magnetic Vortex of the Power from Electrostatic Charges collects electric charges.

The DEC technology is based on the spinning of metallic pieces to produce magnetic fields. The magnetic field or B field is a field that produces a force on a moving charge. This is not the only definition, but it is a useful starting point for operational definitions. It is significant to remember that there are electric fields around every piece of matter in the universe. From atoms to terrestrial bodies, including, yes, the DEC.

Maxwell's equations are used to represent magnetic and electric fields. They show how charges and currents affect them as well as how charges and currents affect the fields. To further comprehend Maxwell's equations, five statements are made.

1. Electric fields are created by charges
2. Electromagnetic waves are produced by accelerating charges.
3. Magnetic fields are created by moving charges.
4. Magnetic fields exert forces on charges traveling perpendicular to the field.
5. An electric field pushes forces against charges.

Therefore, it should go without saying that comprehending the DEC and how it is supposed to work requires knowledge of both the force law and Maxwell's equations. The relevant laws are summarized
as shown;

$$\vec{\nabla} \cdot \vec{E} = \frac{1}{\varepsilon_0} \rho$$

Divergence

Gauss' law

$$\oint \vec{E} \cdot \hat{n} \, dA = \frac{1}{\varepsilon_0} \sum Q_{enclosed}$$

Enclosed flux

$$\vec{\nabla} \cdot \vec{B} = 0$$

Gauss' law (Magnetism)

$$\oint \vec{B} \cdot \hat{n} \, dA = 0$$

$$\vec{\nabla} \times \vec{E} = -\frac{\partial \vec{B}}{\partial t}$$

curl

Faraday's law

$$\oint \vec{E} \cdot d\vec{l} = -\frac{d}{dt} \int \vec{B} \cdot \hat{n} \, dA$$

Circulation

$$\vec{\nabla} \times \vec{B} = \mu_0 \left(\vec{J} + \varepsilon_0 \frac{\partial \vec{E}}{\partial t} \right)$$

Ampere's law

$$\oint \vec{B} \cdot d\vec{l} = \mu_0 \left[\sum I_{end} + \varepsilon_0 \frac{d}{dt} \int \vec{E} \cdot \hat{n} \, dA \right]$$

18

Lorentz Force Law:

$$\vec{F} = q\vec{E} + q\vec{V} \times \vec{B}$$

In example, for divergence of a vector id determines as :

Divergence of a vector example

$$\nabla \cdot \vec{E} = \frac{\rho}{\varepsilon_0} \Rightarrow \oint \vec{E} \, d\vec{A} = \frac{q_{encl}}{\varepsilon_0}$$

$$\vec{E} = k \frac{q}{R^2} \hat{x}$$

$$\nabla \cdot \vec{E} = \frac{\partial E_x}{\partial x} + \frac{\partial E_y}{\partial y} + \frac{\partial E_z}{\partial z}$$

$$\nabla \cdot \vec{E} = \frac{\partial}{\partial x}\left(\frac{kQ}{\partial x}\right) = \frac{\partial}{\partial x}(kQ\,x^{-2}) = kQ\frac{\partial}{\partial x}(x^{-2})$$

$$= -2kQ\,x^{-3} = \frac{-2kQ}{x^3}$$

19

In example given, for a definition of a curl in cylindrical coordinates as :

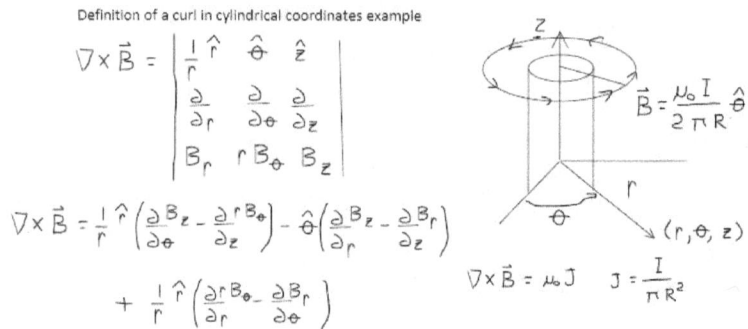

Definition of a curl in cylindrical coordinates example

$$\nabla \times \vec{B} = \begin{vmatrix} \frac{1}{r}\hat{r} & \hat{\theta} & \hat{z} \\ \frac{\partial}{\partial r} & \frac{\partial}{\partial \theta} & \frac{\partial}{\partial z} \\ B_r & r B_\theta & B_z \end{vmatrix}$$

$$\nabla \times \vec{B} = \frac{1}{r}\hat{r}\left(\frac{\partial B_z}{\partial \theta} - \frac{\partial r B_\theta}{\partial z}\right) - \hat{\theta}\left(\frac{\partial B_z}{\partial r} - \frac{\partial B_r}{\partial z}\right)$$

$$+ \frac{1}{r}\hat{r}\left(\frac{\partial r B_\theta}{\partial r} - \frac{\partial B_r}{\partial \theta}\right)$$

$$\vec{B} = \frac{\mu_o I}{2\pi R}\hat{\theta}$$

$$\nabla \times \vec{B} = \mu_o J \qquad J = \frac{I}{\pi R^2}$$

(r, θ, z)

Gravitoelectromagnetism equations

Gravitoelectromagnetic (GEM) equations are based on general relativity theories, which state that, in a specific limiting scenario, equations with the same form as in classical electromagnetism can describe the gravitational field created by a rotating object (or any rotating mass-energy). Starting from the basic equation of general relativity, the Einstein field equation, and assuming a weak gravitational field or reasonably flat spacetime, the gravitational analogs to Maxwell's equations for electromagnetism, called the "GEM

20

equations", can be derived. GEM equations compared to Maxwell's equations are

GEM equations	Maxwell's equations
$\nabla \cdot E_g = -4\pi G \rho_g$	$\nabla \cdot E = \dfrac{\rho}{\varepsilon_0}$
$\nabla \cdot B_g = 0$	$\nabla \cdot B = 0$
$\nabla \times E_g = -\dfrac{\partial B_g}{\partial t}$	$\nabla \times E = -\dfrac{\partial B}{\partial t}$
$\nabla \times B_g = -\dfrac{4\pi G}{c^2} J_g + \dfrac{1}{c^2}\dfrac{\partial E_g}{\partial t}$	$\nabla \times B =$

Where:

E_g is a gravitoelectric field (conventional gravitational field) with SI unit ms^{-2}
E is the electric field
B_g is a gravitomagnetic field, with SI unit s^{-1}
B is the magnetic field
Rho $_G$ is mass density with SI unit $kg\ m^{-3}$
Rho is charge density
J_g is mass current density or mass flux (J_g = Rho$_g$ v_p is the velocity of the mass flow) with SI unit $kg\ m^{-2}\ s^{-1}$
J is electric current density
G is a gravitational constant
Epsilon $_0$ is a vacuum permittivity
c as both a speed of propagation of gravity and the speed of light

This QuickSheet illustrates vector derivative operators in both numerics and symbolics, relative to rectangular x-y-z coordinates. Note that we have capitalized Grad, Div and Curl, which is essential here to distinguish these names from internal names used by the symbolic processor.

In example given: for reference Illustrates vector derivative operators in both numeric and symbolics, relative to rectangular XYZ coordinates.

Symbolic evaluation:

$$\text{Grad}(f, x, y, z) \rightarrow \begin{bmatrix} 2 \cdot x \cdot y \cdot z^3 \\ x^2 \cdot z^3 \\ 3 \cdot x^2 \cdot y \cdot z^2 \end{bmatrix}$$

Numeric evaluation:

$$\text{Grad}(f,1,1,1) = \begin{bmatrix} 2 \\ 1 \\ 3 \end{bmatrix}$$

If Grad=∇, then Div=$\nabla \cdot A$.

$$\text{Div}(A,x,y,z) := \frac{d}{dx}\left[\text{tr}\left[\left(A(x,y,z)^T\right)^{<0>}\right]\right] + \frac{d}{dy}\left[\text{tr}\left[\left(A(x,y,z)^T\right)^{<1>}\right]\right] + \frac{d}{dz}\left[\text{tr}\left[\left(A(x,y,z)^T\right)^{<2>}\right]\right]$$

Symbolic evaluation:

$$\text{Div}(A,x,y,z) \rightarrow$$

Numeric evaluation:

$$\text{Div}(A,1,1,1) =$$

If Grad=∇, then Curl=$\nabla \times A$.

$$\text{Curl}(A,x,y,z) := \begin{bmatrix} \frac{d}{dy}\left[\text{tr}\left[\left(A(x,y,z)^T\right)^{<2>}\right]\right] - \frac{d}{dz}\left[\text{tr}\left[\left(A(x,y,z)^T\right)^{<1>}\right]\right] \\ \frac{d}{dz}\left[\text{tr}\left[\left(A(x,y,z)^T\right)^{<0>}\right]\right] - \frac{d}{dx}\left[\text{tr}\left[\left(A(x,y,z)^T\right)^{<2>}\right]\right] \\ \frac{d}{dx}\left[\text{tr}\left[\left(A(x,y,z)^T\right)^{<1>}\right]\right] - \frac{d}{dy}\left[\text{tr}\left[\left(A(x,y,z)^T\right)^{<0>}\right]\right] \end{bmatrix}$$

Lorentz force

For a test particle whose mass m is "small", in a stationary system, the net (Lorentz) force acting on it due to a GEM field is described by the following GEM analog to the Lorentz force equation:

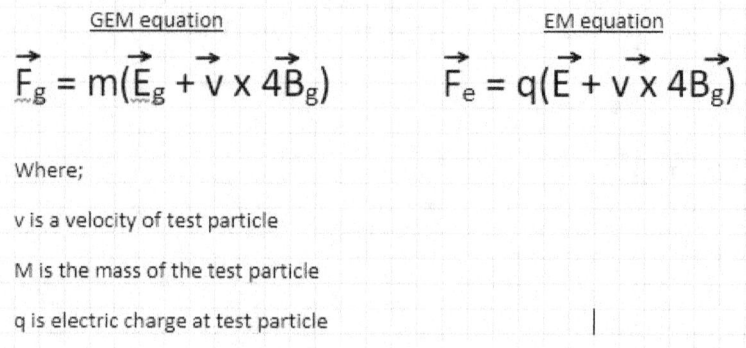

GEM equation

$$\vec{F}_g = m(\vec{E}_g + \vec{v} \times 4\vec{B}_g)$$

EM equation

$$\vec{F}_e = q(\vec{E} + \vec{v} \times 4\vec{B}_g)$$

Where;

v is a velocity of test particle

M is the mass of the test particle

q is electric charge at test particle

Rotational Dynamics

Since the foundation of the DEC is based on the spinning of metallic components to produce magnetic fields, then rotational dynamics plays an important role in the understanding of torque and momentum relationship of the DEC.

Torque is the time derivative of angular momentum as force is the time derivative of linear momentum

$$\frac{d}{dt}L = \tau_{ext} \quad \text{is parallel to} \quad \frac{d}{dt}P = F_{ext}$$

Now look to the kinetic energy (KE) associated with rotating particles. Each particle or peace of a rotating mass has a linear speed $v = r\omega$, and the KE is a sum

$$KE = \frac{1}{2}(m_1v_1^2 + m_2v_2^2 + \ldots) = \frac{1}{2}(m_1r_1^2 + m_2r_2^2 + \ldots)\omega^2$$

The sum of mv^2 for the collection of pieces is called the rotational inertia I, so that the KE can be written compactly as

$$KE = \frac{1}{2}I\omega^2$$

The rotational inertia I, for reference of a regenerative accumulator is illustrated by example;

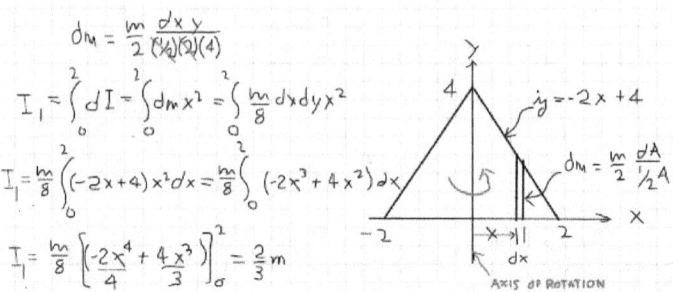

$$dm = \frac{m}{2} \frac{dx\, y}{(\frac{1}{2})(2)(4)}$$

$$I_1 = \int_0^2 dI = \int_0^2 dm\, x^2 = \int_0^2 \frac{m}{8} dx\, dy\, x^2$$

$$I_1 = \frac{m}{8} \int_0^2 (-2x+4) x^2 dx = \frac{m}{8} \int_0^2 (-2x^3 + 4x^2)\, dx$$

$$I_1 = \frac{m}{8} \left[\frac{-2x^4}{4} + \frac{4x^3}{3} \right]_0^2 = \frac{2}{3} m$$

$y = -2x + 4$

$$dm = \frac{m}{2} \frac{dA}{\frac{1}{2}A}$$

AXIS OF ROTATION

The total momentum of a rotating mass point is mvr. If the mass is rotating, then v is perpendicular to r leading to a simple statement of angular momentum.

Angular momentum is

$$L = mvr = mr^2\, \omega = I\omega$$

By definition

$$\tau = \frac{d}{dt} L \quad = \quad I \cdot \frac{d}{dt} \omega \quad = \quad I\alpha$$

The power is

$$\frac{d}{dt}KE = \frac{d}{d\blacksquare}\left(\frac{1}{2}\,I\omega^{2}\right) = I\omega\cdot\frac{d}{dt}\omega \quad . = \quad I\omega\alpha = \tau\omega$$

The Central Electrical System

The Central Electric System includes a set of batteries, mounted inside the device initially feeds the entire electrical system of the craft consists of the electrical connection of countless electrical cells of any known type (lead-acid, nickel-cadmium, metal hydride, etc.). Standard electrical circuits and microprocessors are used for speed and directional control of DC motors. Again, when the system reaches a minimal rotation that allows the rotor and the stator to work as a generator, switching the electromagnets allows energy regeneration (charging) of the batteries.

Regenerative Accumulator

The regenerative accumulator, attached to the internal rotating platform is the power core of the craft described as a device composed of an electrically isolating or conductor material with a biconic geometric shape that rotates inside a magnetic field. It is also referred to as a battery because of its hollow, spherical interior, which houses the electrolyte. A regenerative accumulator is shown;

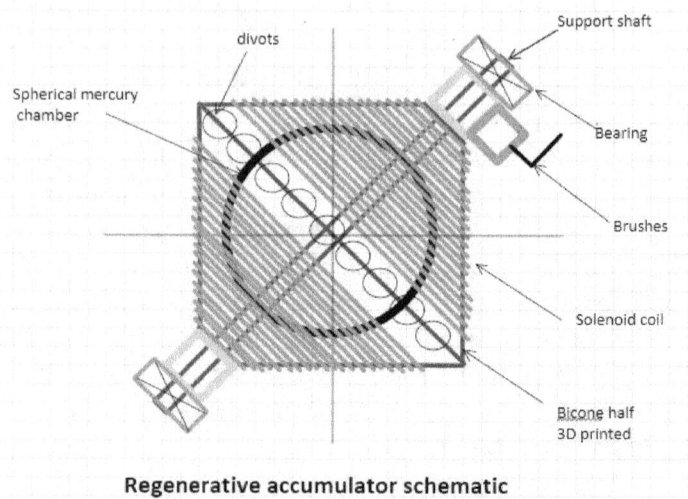

Regenerative accumulator schematic

The battery rotates in this magnetic field. It's the union of two 3-D printed presto-conical sections, two right angle sections. Work in a sense like a turbine reactive channel where atmosphere a flow of air and compressed air aids in rotation. The battery of the regenerative accumulator (bicone with metallic electrolyte) is surrounded by a magnetic field produced by a biconical shaped coil. As the coil rotates with the bicone, it must be

28

energized by applied pulses (through brushes or otherwise) placed at the apex of the upper and lower cones. When the internal platform rotates, the air pressure on the bicone surfaces, also sets it in free rotation, counter-clockwise within the magnetic field and, due to its circular movement, generates electricity.

Here it is the same as in the Faraday unipolar generator, where the magnetic field acts on a rotating electrically conductive material producing electrical energy.

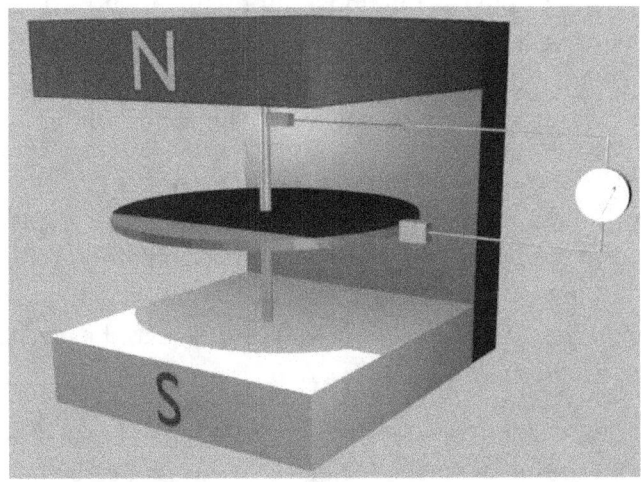

The expression "using the straight lines of magnetic pressure force with motion" means that the vector of the magnetic field is perpendicular to the vector of the linear speed (tangent) of the bicone. The "eddy currents" are produced by the Lorentz force when the bicone rotates within this magnetic field, that is, the electrons in the electrolyte (metallic mercury) conductor layer are displaced to the periphery and the rotation of the device produces an electric current.

The expression "uni-polarized" refers to the unipolar magnetic field, different from the multipolar motors/generators referred to in "looped rotating circuits". The expression "actively repelled by the mass" is due to the phenomenon of the Lorentz force that pushes electric charges towards the bicone equator.

The initial magnetic field is produced by the winding with biconical shape that starts at the apex of one cone and ends at the apex of the other. When the accumulator is in rotation, a difference in electric potential is produced between the center and the periphery (equator) of the bicone due to the electric charge's separation.

The electric current, produced by the rotation of these charges, reinforces the initial magnetic field, so a series of turbine channel divots on the bicones must be fixed such that the rotation produces a magnetic field in the right direction (defined by the right-hand rule), that reinforces the original magnetic field. If the rotation is sufficient, the device enters positive feedback.

Metallic mercury (Hg), which was specifically employed by Faraday in his first unipolar engine experiment in 1821, is the electrolyte used in the hollow spherical interior of the bicone for the reasons proposed for the accumulator. A spherical Marconi dynamo based on the rotation (vortex) of metallic mercury is also mentioned, and given that Marconi was a student of Nikola Tesla, it is reasonable to assume that Tesla provided him with the information.

Accordingly, both Dr. Korovyakov's unipolar electric motor and Marconi's dynamo, which used a fluid mercury rotor, operated on the same principles. After working with Tesla in the 1920s, Marconi created the dynamo. It was an electric dynamo with an outside surface covered in coils and a hollow, spherical stator. Metallic mercury that was partially inserted into the sphere served as a fluid rotor, rotating around a vertical axis (or any other axis).

The accumulator's coil

The coil used to create the initial magnetic field on the accumulator bicone does not have a ferromagnetic core, but the surface density of the magnetic charge $B = u_0 H$ must be high enough to separate the electrons from the conducting layer of the metallic mercury inside the bicone. This coil is also used to move the device's platforms as a motor/generator. As a result, an air core coil must have its pulsing electric current calculated. To begin, we can think about producing a magnetic field with a magnetic field density of about 1 MA/m (10^6 A/m), which is roughly similar to a 1.25 T neodymium magnet.

The biconical shape of the coil creates difficulties in the calculation, so as an approximation, we will use the equation for calculating the magnetic field for a cylindrical coil, without introducing magnetic material:

$$N \cdot I_E = H \cdot L \quad \longrightarrow \quad I_E = \frac{H \cdot l}{N} = \frac{B \cdot l}{\mu_o \cdot N}$$

With:

N = Number of turns

I_E = Electric current [A]

H = Magnetic field [A m^{-1}]

B = Surface density of magnetic charge [Wb m^{-2}][T]

μ_o = Magnetic permeability of free space 1.2566*10^{-6} Wb A^{-1} m^{-1}

l = coil length (m)

In example
Biconical coil made of copper tube with 100 turns and 30 centimeters high

$$H := 10^6 \text{ A m}^{-1} \qquad l := .3 \text{ m} \qquad \mu_o := 1.2566 \cdot 10^{-6} \text{ Wb A}^{-1} \cdot \text{m}^{-1} \qquad N := 100$$

$$I_E := \frac{H \cdot l}{N} \qquad I_E = 3 \bullet 10^3 \bullet A$$

$$B := \frac{I_E \cdot \mu_o \cdot N}{l} \qquad B = 1.257 \bullet \text{kg} \bullet \text{s}^{-2} \bullet \text{A}^{-1} \qquad B = 1.257 \, ^\circ\text{T} \qquad B = 1.257 \, ^\circ\text{Wb} \cdot \text{m}^{-2}$$

Electric Charge Separation from the Electrolyte

The mercury vortex can behave as an electric current loop thanks to the second and third principles of electric charge separation. The magnetic field created by activating the coil around the bicone is applied to metallic mercury by the high rotation of the accumulator, which results in an electric current that strengthens the initial magnetic field.

The magnetic field that induces an electric potential in the passage through the air gap of the electromagnets attached on the external platform and recharges the main battery bank increases with rotational speed.

Knowing the number of electric charges (free electrons) in the volume of liquid mercury that circulates per second in the interior vortex of the bicone allows one to determine the electric current that flows inside regenerative accumulators. The maximum electrical current that the accumulator is capable of producing occurs in the maximum separation situation, which occurs when all of the electrons in the final electronic layer of the electrolyte atoms are revolving at the edge of the bicone equator.

34

Applying the equivalent electric charge formula, we have:

$$q_E = n_e \cdot e \cdot S \cdot l = n_e \cdot e \cdot V$$

where;

q_E = Electric Charge C

N_E = Volumetric density of electric charges [electron m^{-3}]

E = Electric charge of electron = $1.602 * 10^{-19}$ C

V = Volume of mercury = $4/3 * \pi r^3$ m^3

Applying the equivalent electric current formula, we have:

$$I_E = q_e \cdot f = n_e \cdot e \cdot V \cdot f$$

where;

I_E = Electric Current [I]

q_E = Electric Charge [C]

f = Frequency of the rotation = V_{RPM} / 60 [cycle s^{-1}] [Hz]

The magnetic field will be applied to the entire core region of the accumulators, and the magnetic field in the outermost regions will be proportional to the electric currents present inside those regions' diameters. When we use the magnetic field formula without adding magnetic material, we

35

get:

$$H = \frac{I_E}{2r} \quad \Rightarrow \quad B = \mu_0 H \; .$$

With:

H = Magnetic field [A m^{-1}];
B = Surface density of magnetic charge [Wb m^{-2}] [T];
μ_0 = Magnetic permeability of free space = 1.2566*10^{-6} Wb A^{-1} m^{-1};
I$_E$ = Electric current [A];
r = Average internal radius (of the electrolyte) [m].

The concentration of ions n_e may be calculated from the density of mercury Hg = 13.58 g/cm^3 and its atomic mass M_{Hg} = 200.59 g/mol. In the case of mercury, this concentration is equal to the concentration of free electrons because each atom contributes an electron. With an atomic mass of one mole of atoms and an Avogadro NA = 6.022x10^{23} atom/mol, we may calculate:

$$n_e = \rho_{Hg} \frac{N_A}{M_{Hg}} = 13.58 \frac{6.022*10^{23}}{200.59} = 4.077*10^{28} \; electron \, m^{-3} \; .$$

In example, an electrical insulating material (nylon, celeron, bakelite, etc.) accumulator with a 25 cm diameter and 25 cm height has a hollow, 20 cm diameter spherical core where metallic mercury is deposited until it completely fills the space.

At 12,000 RPM, the accumulator spins.

Applying the equivalent electric charge formula, we have:

$$q_E = n_e \, e \, V = 4.077 * 10^{28} * 1.602 * 10^{-19} * 4.189 * 10^{-3} = 2.736 * 10^7 \, A$$

With:

I_E = Electric current [A];
n_e = 4.077*10^{28} electron m^{-3};
e = 1.602*10^{-19} C;
$V = 4/3 * \pi r^3 = 4/3 * \pi (0.1)^3 = 4.189 * 10^{-3}$ m^3.

Applying the equivalent electric current formula, we have:

$$I_E = q_E f = 2.736 * 10^7 * 200 = 5.472 * 10^8 \, A$$

With:

I_E = Electric current [A];
q_E = 2.736*10^7 C;

$f = v_{RPM}/60 = 12,000/60 = 200$ Hz.

Applying the formula of the magnetic field without the introduction of magnetic material, we have:

$$H = \frac{I_E}{2r} = \frac{5.472 * 10^8}{2 * 0.05} = 5.472 * 10^9 \, A m^{-1} \quad .$$

With:

H = Magnetic field [A m^{-1}];
I_E = 5.472*10^8 A;
r = 0.05 m.

$$B = \mu_0 H = 1.2566 * 10^{-6} * 5.472 * 10^9 = 6.876 * 10^3 T \quad .$$

With:

B = Surface density of magnetic charge [Wb m^{-2}] [T];
μ_0 = 1.2566*10^{-6} Wb A^{-1} m^{-1};
H = 5.472*10^9 A m^{-1}.

In practice, only a portion of this separation will be accomplished, which is determined by Lorentz's Force F=q$_E$ ($\vec{v}$ x$\vec{B}$) created by the rotation of the electrolyte within the magnetic field of the coil. An electric field is created when electric charges separate $\vec{E}=\vec{v} \times \vec{B}$ is possible to determine how many electric charges are separated from the electrolyte using the Hall effect, a phenomenon.

We can calculate the quantity of electric charges and the electric current as a function of the rotation of the electrolyte because the electric field is produced by the distribution of electric charges, just like in capacitors.

A radial electric field perpendicular to the axis of rotation is produced when the separation of electric charges in the electrolyte occurs in a direction that is both perpendicular to the applied magnetic field and to the rotation of the bicone.

The distribution of electric charges in this circumstance can be roughly compared to that of a capacitor, whose average surface is defined by height (half the sphere's diameter) and length (half the sphere's perimeter).

The following equations are revealed to approximate the following:

$$\vec{D} = \varepsilon \vec{E} \quad \Rightarrow \quad D = \frac{q_E}{S} = \varepsilon E = \varepsilon \frac{V_E}{l} \quad \Rightarrow \quad q_E = \varepsilon S E = \varepsilon \frac{S}{l} V_E = C_E V_E \quad ;$$

$$\vec{E} = \vec{v} \times \vec{B} \quad \Rightarrow \quad q_E = \varepsilon S E = \varepsilon S v B \quad .$$

With:

q_E = Electric charge [C];
D = Surface density of electric charge [C m^{-2}];
ε = Electric permittivity [C V^{-1} m^{-1}] [F m^{-1}];
E = Electric field [V m^{-1}];
V_E = Electric potential [V];
C_E = Electric capacitance [F];
B = Surface density of magnetic charge [Wb m^{-2}];
v = Average linear velocity of electrolyte = $2\pi r/2 * f = \pi r \, v_{RPM}/60$ [m s^{-1}];
S = Average surface of capacitor = $r * 2\pi r/2 = \pi r^2$ [m^2];
l = Distance [m].

The above-calculated quantity of electric charges and the electrolyte's rotational frequency together define the amount of

$$I_E = q_E f = \varepsilon S v \, Bf \quad.$$

With:

I_E = Electric current [A];
q_E = Electric charge [C];
f = Frequency of rotation = $v_{RPM}/60$ [cycle s^{-1}] [Hz].

electric current:

In example, an accumulator made with electrical insulating material (nylon, celeron, bakelite etc.) with 25 cm in diameter and 25 cm in height, it has a hollow spherical center of 20 cm in diameter where metallic mercury is deposited until it fills its volume. The accumulator spin at 12,000 RPM and the applied magnetics is 1 MA/m (1.25 T).

Applying the equivalent electric charge formula, we have:

$$q_E = \varepsilon S v \, B = 1.2566 * 10^{-6} * 3.1416 * 10^{-2} * 6.2832 * 10^{1} * 1.25 = 3.100 * 10^{-6} C$$

With:

q_E = Electric charge [C];
ε = 1.2566*10^{-6} C V^{-1} m^{-1} [F m^{-1}];
B = 1.25 Wb m^{-2};
v = $\pi r \, v_{RPM}/60$ = 6.2832*10^{1} m s^{-1};
S = πr^2 = 3.1416*10^{-2} m^2.

We see that the amount of electric charge separated in the electrolyte is very small, compared to the condition of total charge separation.

Applying the equivalent electric current formula, we have:

$$I_E = q_E f = 3.100*10^{-6}*200 = 6.200*10^{-4} A \ .$$
With:
I_E = Electric current [A];
$q_E = 3.100*10^{-6}$ C;
$f = v_{RPM}/60 = 200$ cycle s^{-1} [Hz].

Since the electric current is so little, even if the magnetic field and rotation were multiplied by a factor of 10, it would still be insufficient to meet the demands of an increased initial magnetic field. There are two possibilities: the first is that the accumulator's operation is independent of this charge separation from the electrolyte, and the second is that the coil is producing a magnetic field that is significantly stronger than that predicted to efficiently separate a much higher number of electrical charges from the electrolyte.

Electrostatic Charges Collected from the Atmosphere

The accumulator's revolving magnetic fields displace electrostatic charges from the atmosphere and project them outward, creating an electric current that simultaneously strengthens the initial magnetic field. Because the magnetic field always projects these charges to the periphery, the density of electrostatic charge drops close to the accumulator.

However, fresh charges from the atmosphere travel to maintain density, and the cycle continues as long as the accumulator rotates. Each of the six accumulators, which are dispersed across the internal platform, has a magnetic vortex that draws electrical charges from the atmosphere and directs them onto the capacitor plates.

These charges are projected by the equator of the accumulators due to their biconic design, reaching the top and lower portions of the capacitor plates on the right and left of each accumulator, respectively. As a result, each capacitor plate receives an equal number of electrostatic charges, which when they build up significantly ionize the plates. With the platform's rotation comes a large electric current and, consequently, a high magnetic field in the spin center of the platforms.

Additionally, the magnetic fields surrounding the spin center are moved by the rotation of the inner platform, where the accumulator is situated, and a significant number of electrostatic charges from the environment surrounding the disk are also moved. In example, an accumulator with 25 cm in diameter and 25 cm high, has an initial magnetic field H = 1 MA/m applied by its coil. The accumulator spin speed is 12,000 RPM.

The volume of the cylinder minus the volume of the bicone will estimate the approximate quantity of electrostatic charge influenced by the magnetic field:

$q_E = n_e e V = 4 * 10^{25} * 1.602 * 10^{-19} * 8.181 * 10^{-3} = 5.243 * 10^4 C$.

With:

q_E = Electric charge [C];

n_e = Volumetric density of electric charge of the atmosphere = $4 * 10^{25}$ electron m^{-3};

$e = 1.602 * 10^{-19}$ C;

V = Volume = $2/3 * \pi r^2 * h = 2/3 * \pi * 0.125^2 * 0.25 = 8.181 * 10^{-3} m^3$.

The initial surface density of magnetic charge is:

$B = \mu_0 H = 1.2566 * 10^{-6} * 10^6 = 1.2566 T$.

With:

B = Surface density of magnetic charge [Wb m^{-2}] [T];

$\mu_0 = 1.2566 * 10^{-6}$ Wb $A^{-1} m^{-1}$;

$H = 10^6$ A m^{-1}.

The velocity of atmospheric air inside the magnetic field of the accumulator coil is calculated using the average linear speed of the bicone rotation at the average radius:

$$\omega = \frac{2\pi}{60} V_{RPM} \quad \Rightarrow \quad \bar{v} = \omega \frac{r}{2} = \frac{2\pi}{60} V_{RPM} \frac{r}{2} = \frac{\pi}{60} V_{RPM} r = \frac{\pi}{60} 12,000 * 0.125 = 78.54\, m\, s^{-1}$$.

With:

v = Average velocity [m s^{-1}];

V_{RPM} = Spin velocity = 12,000 RPM;

r = bicone radius = 0.125 m.

The force on the electrostatic charge is:

$$F = e\,v\,B = 1.602 * 10^{-19} * 78.54 * 1.2566 = 1.579 * 10^{-17}\,N$$

With:

F = Force [N];
e = $1.602 * 10^{-19}$ C;
v = 78.54 m s^{-1};
B = 1.2566 T.

Charge acceleration is defined as a function of mass:

$$a = \frac{F}{m_e} = \frac{1.579 * 10^{-17}}{9.109 * 10^{-31}} = 1.733 * 10^{13}\,m\,s^{-2} \quad .$$

With:

a = Acceleration [m s^{-2}];
F = $1.579 * 10^{-17}$ N;
m_e = Mass of electron = $9.109 * 10^{-31}$ kg.

$$t_1 = \sqrt{\frac{l_m}{a}} = \sqrt{\frac{0.0625}{1.733 * 10^{13}}} = 6.01 * 10^{-8}\,s \quad .$$

With:

t_1 = Acceleration time [s];
l_m = Magnetic field length (half radius) = 0.0625 m;
a = $1.733 * 10^{13}$ m s^{-1}.

Charges velocity after acceleration:

45

$$v_o = \sqrt{a \, l_m} = \sqrt{1.733*10^{13}*0.0625} = 1.04*10^6 m\,s^{-1} \quad .$$

$$t_2 = \frac{d_2}{v_o} = \frac{0.5}{1.04*10^6} = 4.81*10^{-7} s \quad .$$

With:

t_2 = Constant velocity time [m s^{-1}];
d_2 = Distance between Utron and capacitor plate ≈ 0.5 m;
v_o = 1.04*10^6 m s^{-1}.

The average time of the charge passage from the magnetic field's center to the capacitor plates
is:

$$t = t_1 + t_2 = 6.01*10^{-8} + 4.81*10^{-7} = 5.41*10^{-7} s \quad .$$

The amount of electric charges that each capacitor plate receives per secc

$$I_E = \frac{q_E}{t} = \frac{5.243*10^4}{5.41*10^{-7}} = 9.69*10^{10} C\,s^{-1} \quad .$$

This is merely an estimate of the volume of electric charges that a tiny accumulator unit with an initial magnetic field similar to a conventional magnet will project. The reason the plates are referred to as "capacitor plates" is obvious: they store a lot of electric charge. Because of how much this magnetic field is magnified when in use, this gadget is ideal as an energy generator.

External Magnetic Field

There are two components to the peripheral magnetic system:

1. Six Accumulator regenerating accumulators that rotate in counter-rotation with their axis of rotation at 45° from the vertical and are spaced 60° apart around the perimeter of the internal platform that rotates in direct rotation;
2. A group of twelve electromagnets spaced at 30° intervals around the circumference of the external structure of the disc, which rotates in counter-rotation near the craft's hull.

Figure 7 Three D view looking down showing internal,external and stationary platform assembly

The twelve electromagnets fixed to the hull's external construction have the shape of a horseshoe or the "C" cores of transformers. A coil of enameled copper wire is twisted around the middle leg of each core, and the accumulators pass through the core air gap, inducing an electric potential in the electromagnetic coils due to their magnetic fields.

Because only six of the twelve electromagnets are powered at the same time during the passage of the six accumulators, we have two circuits that are 30° apart. At first, the system serves as a motor, but after a certain spin, it transforms into a self-sustaining generator, giving energy to recharge the central battery. Because of the design of the accumulators, only a fraction of the magnetic field produced by them is passed to the core of the electromagnets in the perimeter when the system operates as a generator. This is important because its magnetic fields become extremely powerful, potentially saturating the core.

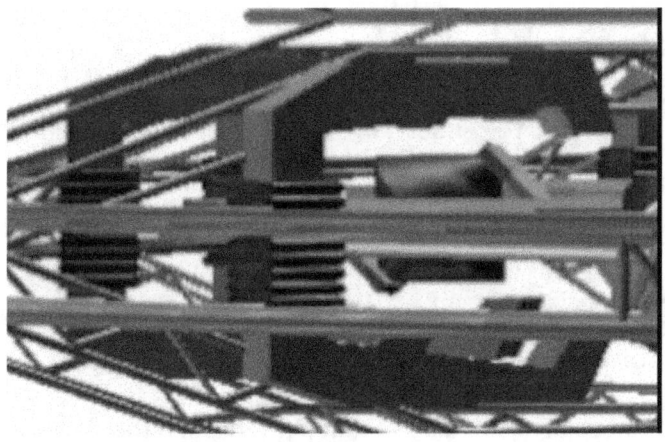

Figure 8 Solid model showing c shaped transformer coil and accumulator

There is a sinusoidal fluctuation in the surface density of the magnetic charge of the core when the accumulator passes over the air gap of the core, and this variation (called magnetic induction) generates an electric potential that is inversely proportional to its duration. Each coil's electric potential will increase proportionally to the number of turns:

$$f = 1/t \quad , \quad q_M = BS \quad \Rightarrow \quad V_E = -N \frac{dq_M}{dt} = -NS \frac{dB}{dt} = -NSBf \quad .$$

With:

V_E = Electric potential [V];
N = Number of coil turns;

q_M = Magnetic charge [Wb];
B = Surface density of magnetic charge [Wb m^{-2}] [T];
S = Core section area [m^2];
t = Time of senoidal wave [s];
f = Frequency of senoidal wave [cycles s^{-1}] [Hz].

The number of accumulators that pass through the ferromagnetic core will be doubled by the frequency of the sine wave because each accumulator that does so creates an electric potential in the coil:

$$f = N_r \frac{v_{RPM}}{60} \quad .$$

With:

f = Frequency of induced sinusoidal electrical potential [Hz];

N_r = Number of accumulators;

v_{RPM} = Translation speed of rollers around the plate [RPM].

As an example, in each coil, with 6 accumulators surrounding the platform, it will be induced a sine wave of frequency equivalent to:

$$f = 6 * \frac{v_{RPM}}{60} = \frac{v_{RPM}}{10} \; Hz \quad .$$

The energy that can be retrieved is equal to the energy of the magnetic field and depends on the magnetic energy density and magnetic volume of each

51

accumulator:

$$U=\frac{1}{2}BHSd=\frac{1}{2}\frac{B^2}{\mu}Sd \quad .$$

With:

    U = Energy [J];
    B = Surface density of magnetic charge of the accumulator [Wb m^{-2}] [T];
    H = Magnetic field intensity of the accumulator [A m^{-1}];
    μ = Magnetic permeability of the accumulator [Wb A^{-1} m^{-1}] [H m^{-1}];
    S = Magnetic surface of the accumulator [m^2];
    d = High of the accumulator [m].

The electrical power that can be extracted from the group of accumulators in one coil is defined by a frequency that is determined by the passage of each accumulator through the peripheral
coils:

$$P=Uf=\frac{1}{2}\frac{B^2}{\mu}Sdf \quad .$$

With:

    P = Power [W];
    U = Energy [J];
    f = Frequency of induced sinusoidal electrical potential [Hz].

We will have 12 times this power if the device contains 12 coils. Each coil is calculated using the standard method for calculating transformers. Transformers can be calculated using the following formula:

$$N = \frac{V_E}{4,44\, B_{MAX}\, S f}$$

With:

N = Number of coil turns;
V_E = Electric potential (RMS) applied to coil [V];
B_{MAX} = Maximum surface density of magnetic charge of ferromagnetic core [Wb m^{-2}] [T];
S = Core section area [m^2];
f = Operating frequency [Hz].

In example, with a diameter of 1 m and a height of 1 m, six accumulators made of electrical insulating material (such as nylon, celeron, bakelite, etc.) have a hollow, spherical center with a 0.8 m diameter where metallic mercury is deposited until it fills the space.

On the outside, a complex coil of copper wire that has been enameled is wound from the apex of one cone to the apex of the other. Each accumulator's final spin is 12,000 RPM.

In the outer perimeter, 12 "C"-shaped electromagnets built of ferromagnetic silicon steel sheets with a relative magnetic permeability of 4,000 can sustain 1.2 T [Wb/m^2] of surface magnetic charge density. Their air gaps are 20 x 20 cm square and have a height of 1 m. There will be gaps of 15 cm above and

53

below when the accumulators pass through the air gap in this scenario.

The accumulators' internal platform and the electromagnets' external platform rotate in opposite directions at 24,000 RPM. We have two 30° lagged circuits that act as a motor and, once they reach self-sustaining speed, as a generator because each coil's output voltage is 12,000 volts, and only six electromagnets are active at a moment in the passage of the six accumulators. To power the disc's center battery, the coils are connected in parallel. The number of electric charges (free electrons) of the volume of liquid mercury that circulate per second is used to compute the electric current that flows inside regenerative accumulators.

Applying the equivalent electric current formula, we have:

$$I_E = n_e \, e \, V f = 4.077 * 10^{28} * 1.602 * 10^{-19} * 8.533 * 10^{-2} * 200 = 1.115 * 10^{11} \, A$$

With:

I_E = Electric current [A];
n_e = Volumetric density of electric charge = $4.077 * 10^{28}$ electron m^{-3};
e = Electric charge of electron = $1.602 * 10^{-19}$ C;
V = Volume of mercury = $4/3 * \pi r^3 = 4/3 * \pi (0.4)^3 = 8.533 * 10^{-2}$ m^3;
f = Frequency of rotation = $v_{RPM}/60 = 12,000/60 = 200$ Hz.

The magnetic field will be applied to the entire middle region of the accumulators and will be proportionate to the electric currents that are present in those regions' outermost regions. When we use the magnetic field formula without

adding magnetic material, we get:

$$H = \frac{I_E}{2r} = \frac{1.115 * 10^{11}}{2 * 0.5} = 1.115 * 10^{11} \, A \, m^{-1} \quad .$$

With:

$\qquad$ H = Magnetic field [A m⁻¹];

$\qquad$ I_E = 1.115*10¹¹ A;
$\qquad$ r = Internal radius = 0.5 m.

$$B = \mu_0 H = 1.2566 * 10^{-6} * 1.115 * 10^{11} = 1.401 * 10^5 \, T \quad .$$

With:

$\qquad$ B = Surface density of magnetic charge [Wb m⁻²] [T];
$\qquad$ μ_0 = 1.2566*10⁻⁶ Wb A⁻¹ m⁻¹;
$\qquad$ H = 1.115*10¹¹ A m⁻¹.

Calculations are made to determine the frequency of the electric potential caused by the passing of six accumulators.

$$f = 6\frac{V_{RPM}}{60} = 6\frac{24{,}000}{60} = 2{,}400\,Hz \quad .$$

With:

> f = Frequency of sinusoidal wave [Hz];
> V_{RPM} = 24,000 RPM.

We will keep in mind that the value of B does not surpass the maximum value permitted by the ferromagnetic cores because only a small portion of the calculated magnetic field would generate electric potential in the electromagnetic coils and the

$$N = \frac{V_{RMS}}{4.44\,B_{MAX}\,Sf} = \frac{12{,}000}{4.44*1.2*4*10^{-2}*2{,}400} = 23.46 \approx 24\,turns$$

With:

> N = Number of coil turns;
> V_{RMS} = 12,000 V;
> B_{MAX} = 1.2 T;
> S = 400 cm² = $4*10^{-2}$ m²;
> f = 2,400 Hz.

15 + 15 cm air gap assures that the core material does not saturate. Calculating coils

The magnetic energy generated by the accumulators can be used to gauge the electrical power provided by the external magnetic system (accumulators plus electromagnets). The ratio of the area of the core section to the area of the accumulators' circumference determines how much magnetic field induces electric potential in electromagnetic coils. The magnetic field tends to be stronger on the axis than on the periphery because the current turns in the liquid metal are located at different distances from the accumulators' axis of rotation. Therefore, we will get an average magnetic field that is less than the anticipated total if the accumulators pass through the air gap in an average position between the axis and the perimeter.

The list of areas is:

1. Core section area = $4*10^{-2}$ m²;
2. Accumulator circumference area = $\pi r 2 = \pi (0.5)^2$
   $= 5.854*10^{-1}$ m².

We'll round up the magnetic field reduction to 100 because the area differences are roughly 20 times apart and the magnetic field isn't as strong as predicted. The following equations provide the amount of energy in the magnetic field that travels through the air gaps:

$$U = \frac{1}{2}\mu_0 H^2 V = \frac{1}{2}1.2566*10^{-6}*|1.115*10^9|^2*2.8*10^{-2}=2.187*10^{10} J \quad .$$

With:

U = Energy [J];
μ_0 = 1.2566*10^{-6} Wb A^{-1} m^{-1};
H = 1.115*10^9 A m^{-1};
V = Magnetic field volume = A*l = 4*10^{-2} * 0.7 = 2.8*10^{-2} m^3.

In the passage of the six accumulators, the magnetic field's available electrical power that may be collected from each coil is as follows:

$$P = U f = 2.187 * 10^{10} * 2,400 = 5.249 * 10^{13} W$$
With:

    P = Electrical power [W];
    U = 2.187*10^{10} J;
    f = 6*24,000/60 = 2,400 Hz.

Due to restrictions on the materials employed, all of this power will not be usable in electromagnets, but it is a power that might be used in the development of magnetic materials with extremely high permeability.

Capacitor plate magnetic propulsion system

The capacitor plate system's goal is to use so-called ambient electricity to store a large number of electrical particles.

When the plates are functioning properly, they ionize and emit a faint bluish-green glow that resembles a corona. The capacitor plates' ionization creates a magnificent glow with a delicate luminous light. It would resemble an electric arc in welding or have a blue-green coloration.

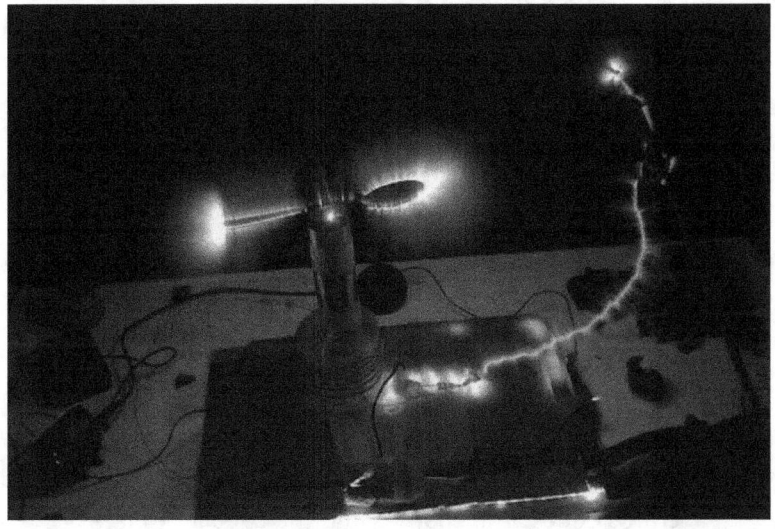

Figure 9 View showing Ionize plates emitting a faint bluish-green glow that resembles a corona.

The electrical charges and ions that are projected from the craft's perimeter into the atmosphere ionize the surrounding air, causing it to glow in a manner akin to that of evacuated tubes due to the low pressure of the ionized air.

As we saw above, ambient electricity charges the capacitor plates, causing them to absorb and accumulate the electrostatic charges that the equipment deflects. The buildup of charges transforms the plates into superconductors and increases the strength of the magnetic field generated on the craft's axis.

Electrostatic charges can be absorbed in two different ways:

1. Electrostatic charges are deflected from the atmosphere and rotated around their circle by each of the six revolving Accumulator generators, causing them to collide with the nearby capacitor plates;

2. Electrostatic charges from the environment are deflected by the vortex-like central magnetic field that is created, and they are then projected outward to the craft's edges. The charges and the capacitor plates collide in this path.

60

The capacitor plates behave like a high-intensity electric current when they are rotating quickly and have excess electric charges, which produces a strong vertical magnetic field on the craft's axis that repels the vertical component of the terrestrial magnetic field and enables the craft to float magnetically in the atmosphere.

In example, a 15-meter-diameter interior platform, a 3-meter-diameter central battery, and six metal capacitor plates folded in "C" configuration to increase their surface area. The capacitor plates are 5 m long, 3 m wide externally, and 2 m wide internally. These plates are made of copper and have a 10 mm thickness. At 12,000 RPM, the platform spins. We will assume that the capacitor plates absorb an additional electrostatic charge that increases the density of free electrons in copper by a factor of 1,000, causing the plates to transition from conducting too superconducting.

Applying the equivalent electric current formula, we have:

$$I_E = n_e e\, S\, l\, f = 8.46*10^{31}*1.602*10^{-19}*150*10^{-2}*200 = 4.066*10^{15}\,A \quad.$$

With:

    I_E = Electric current [A];

    n_e = Volumetric density of electric charge = $8.4538*10^{31}$ electron m^{-3};

    e = Electric charge of electron = $1.602*10^{-19}$ C;

    S = Area of plates = 6 * 2 (5 * 2.5) = 150 m^2;

    l = Plate thickness = 10^{-2} m;

    f = Rotation frequency = $v_{RPM}/60$ = 12,000/60 = 200 Hz.

The middle 5 m radius area will receive 100% of the magnetic field, while the outermost portions will receive a magnetic field proportional to the electric currents inside their diameters. Using the magnetic field formula without the addition of magnetic material, we get:

$$H = \frac{I_E}{2r} = \frac{4.066*10^{15}}{2*2.5} = 8.132*10^{14}\,A\,m^{-1} \quad.$$

With:

    H = Magnetic field [A m^{-1}];

    I_E = Electric current = $4.066*10^{15}$ A;

    r = Internal radius = 2.55 m.

$$B = \mu_0 H = 1.2566*10^{-6}*8.132*10^{14} = 1.022*10^{9}\,T \quad.$$

With:

    B = Surface density of magnetic charge or magnetic induction [Wb m^{-2}] [T];

    μ_0 = Magnetic permeability of atmosphere = $1.2566*10^{-6}$ Wb A^{-1} m^{-1};

    H = $8.132*10^{14}$ A m^{-1}.

The repellent force between the magnetic field formed in the device's center and the vertical component of the terrestrial magnetic field is as
follows:

$$F = q_M H = BSH = 10^{-9} * 19.635 * 8.132 * 10^{14} = 1.597 * 10^7 \, N \, .$$

With:

    F = Repulsion force [N];
    B = Vertical component of terrestrial surface density of magnetic charge = 10^{-9} T;
    S = Area submitted to the magnetic field = $\pi r^2 = \pi(2.5)^2 = 19.635 \, m^2$;

$$H = 8.132 * 10^{14} \, A \, m^{-1}.$$

The amount of gravitational charge (mass) that can be levitated with this force is:

$$q_G = \frac{F}{G} = \frac{1.597 * 10^7}{9.80665} = 1.628 * 10^6 \, kg \, .$$

With:

    q_G = Gravitational charge (mass) [kg];
    F = 1.597*10^7 N;
    G = Terrestrial gravitational field [N kg^{-1}] = g = gravitational acceleration = 9.80665 m s^{-2}.

System of Gravitoinertial Propulsion

Theoretical advancement that allows for the neutralization of gravitational pull in relation to an object's speed by analogy to the balance of forces that keeps a satellite in orbit around the earth.

Calculations were used to quantify the gyroscopic effect, which is a phenomenon in which an object can defy the gravitational pull of the planet while in high rotation. Because the inertial current is the square of the speed, and this creates a gravitational potential, a high-speed external platform can help negate the weight of a discoid device.

The equations are:

$$I_j = v^2 = \omega^2 r^2 = \frac{\omega^2}{2}\left(r_2^2 - r_1^2\right) = V_G \quad \Rightarrow \quad \omega = \sqrt{\frac{2V_G}{r_2^2 - r_1^2}} \quad .$$

With:

  I_j = Inertial current [m² s⁻²];
  ω = Angular speed of object [rad s⁻¹];
  r_1 = Internal radius of object [m];
  r_2 = External radius of object [m];
  V_G = Gravitational potential [m² s⁻²].

The gravitational potential on the planet's equatorial surface is:

$$V_G = k_g \frac{Q_G}{R} = 6.6739 * 10^{-11} \frac{5.976 * 10^{24}}{6.378 * 10^6} = 6.253 * 10^7 \, m^2 s^{-2} \quad .$$

With:

  V_G = Gravitational potential [N m kg⁻¹] [m² s⁻²];
  k_g = Universal gravitational constant = 6.6739*10⁻¹¹ N m² kg⁻² [m³ kg⁻¹ s⁻²];
  Q_G = Gravitational charge (mass) of the Earth = 5.976*10²⁴ kg;
  R = Equatorial radius of Earth = 6.378*10⁶ m.

When the total mass of a discoid craft exceeds the mass of the rotating mass, we must determine the additional gravitational potential by balancing the inertial energy of the rotating gravitational charge's mass against the gravitational energy of the device's total mass:

$$U = q_{G1} V_{G1} = q_{G1} I_{1} = q_{G1} \frac{\omega^2}{2} |r_2^2 - r_1^2| = q_{G2} V_G \quad \Rightarrow \quad \omega = \sqrt{\frac{2V_G}{r_2^2 - r_1^2} \frac{q_{G2}}{q_{G1}}} = \sqrt{\frac{2k_g}{|r_2^2 - r_1^2| R} \frac{Q_G q_{G2}}{q_{G1}}}$$

With:

U = Energy [J];
q_{G1} = Rotating gravitational charge [kg];
q_{G2} = Total Gravitational charge [kg];
V_{G1} = Gravitational potential of the rotating object [m² s⁻²];
V_G = Gravitational potential of the planet [m² s⁻²];
I_1 = Inertial current of the rotating object [m² s⁻²];
ω = Angular speed of the object [rad s⁻¹];
r_1 = Internal radius of platform [m];
r_2 = External radius of platform [m].

In example, a discoid vehicle with a mass of 25,000 kg and an exterior platform that measures 15 meters in diameter is built. Due to the weight of the electromagnets, the exterior platform's mass, which rotates counterclockwise, is 6,000 kg and is dispersed throughout the last meter of the craft's radius. Determine the speeds needed to remove 30% and 100% of the weight in question. In the scenario where we lose 30% of our weight, we have:

66

$$\omega = \sqrt{\frac{2k_g}{\left(r_2^2 - r_1^2\right)R} \frac{Q_G q_{G2}}{q_{G1}}} = \sqrt{\frac{2*6.674*10^{-11}}{\left(15^2 - 14^2\right)6.378*10^6} 0.3 \frac{5.976*10^{24}*25{,}000}{6{,}000}} = 2.322*10^3 \, rad \, s^{-1}$$

$$V_{RPM} = \frac{60}{2\pi}\omega = \frac{60}{2\pi}2.322*10^3 = 2.22*10^4 \, RPM \quad .$$

With:

    V_{RPM} = Rotation speed [RPM];

    q_{G1} = Rotating gravitational charge = 6,000 kg;

    q_{G2} = Total Gravitational charge = 25,000 kg;

    r_1 = Internal radius of platform = 14 m;

    r_2 = External radius of platform = 15 m;

    k_g = Universal gravitational constant = $6.6739*10^{-11}$ N m^2 kg^{-2} [m^3 kg^{-1} s^{-2}];

    Q_G = Gravitational charge (mass) of the Earth = $5.976*10^{24}$ kg;

    R = Equatorial radius of Earth = $6.378*10^6$ m.

In the case of eliminating 100% of the weight we have:

$$\omega = \sqrt{\frac{2k_g}{\left(r_2^2 - r_1^2\right)R} \frac{Q_G q_{G2}}{q_{G1}}} = \sqrt{\frac{2*6.674*10^{-11}}{\left(15^2 - 14^2\right)6.378*10^6} \frac{5.976*10^{24}*25{,}000}{6{,}000}} = 4.239*10^3 \, rad \, s^{-1} \quad .$$

$$V_{RPM} = \frac{60}{2\pi}\omega = \frac{60}{2\pi}4.239*10^4 = 4.05*10^4 \, RPM \quad .$$

System of Artificial Gravity

We are aware that the internal and external platforms rotate counter to one another, keeping the Stationary device motionless. We have two masses rotating in different directions because of the accumulator devices positioned on the internal platform, which are rotating counterclockwise.

The development of a mathematical method for the induction of a gravitational potential with mass rotation allows us to create an artificial gravitational center with two masses rotating in opposite directions.

Six accumulators rotate in counter revolution alongside the capacitor plates on the interior platform, which also rotates in direct rotation. These opposing rotations enable the creation of a gravitational attraction point where the gadget and anything else subject to the gravitational field will be pushed. This point is formed in the axis of rotation of the platform and the accumulators. A less intense gravitational center also arises in the axis of rotation of the interior and external platforms, which rotate in opposite directions. We can refer to it as artificial gravity.

In other words, if one platform has more inertial energy than the other, a gravitational force will be exerted in that direction. The gravitational center is proportional to the quantity of inertial energy of the same amount for opposite rotations. The same equations that have already been used are used to compute the gravitational potential associated with this

gravitational
center:

$$I_I = v^2 = \omega^2 r^2 = \frac{\omega^2}{2}\left(r_2^2 - r_1^2\right) = V_G \quad \Rightarrow \quad U = q_G V_G = q_G I_I = q_G \frac{\omega^2}{2}\left(r_2^2 - r_1^2\right) \; .$$

With:

I_I = Inertial current [m² s⁻²];
ω = Angular velocity of object [rad s⁻¹];
r = Distance from object to spin center [m];
r_1 = Internal radius of platform [m];
r_2 = External radius of platform [m];
V_G = Gravitational potential [m² s⁻²];
U = Energy [J];
q_G = Rotating gravitational charge [kg].

On the other hand, considering what was exposed in the article Gravitational Charge [6], the energy associated with a gravitational field may be expressed by:

$$u = \frac{1}{2}\gamma_0 G^2 \quad \Rightarrow \quad U = \frac{1}{2}\gamma_0 G^2 V \; .$$

With:

u = Volumetric density of energy [J m⁻³];
U = Energy [J];
γ_0 = Gravitational permeability of vacuum = 1.19230*10⁹ [kg² N⁻¹ m⁻²];
G = Gravitational field [N kg⁻¹] = acceleration [m s⁻²];
V = Volume of the gravitational field [m³].

Matching the two energies, we can estimate the value of the artificial gravitational field produced by the rotation of a quantity of gravitational charge (mass):

$$U = q_G \omega r^2 = q_G \frac{\omega^2}{2}\left(r_2^2 - r_1^2\right) = \frac{1}{2}\gamma_0 G^2 V \quad \Rightarrow \quad G = \sqrt{\frac{2U}{\gamma_0 V}} = \sqrt{\frac{2q_G}{\gamma_0 V}\,\omega r^2} = \sqrt{\frac{q_G}{\gamma_0 V}\,\omega^2\left(r_2^2 - r_1^2\right)} \; .$$

In example, a gadget with a 12 m-diameter internal platform and six accumulators evenly spaced at 60° intervals all around the circle. Accumulators are installed in accordance with the original design at a 45° angle to the craft's center axis. Each accumulator weighs 200 kg, has a 50 cm diameter, and has centers that are 4 m from the axis. Its counter-rotation is designed to counteract the platform's 24,000 RPM direct rotation.

The inertial energy of the set is given by:

$$U = q_G \omega r^2 = 1,200 * 2.513 * 10^3 * 4^2 = 1.2125 * 10^{11} J .$$

With:

U = Energy [J];
q_G = 6*200 = 1,200 kg;
$\omega = 2\pi/60 \ v_{RPM} = 2\pi/60 * 24.000 = 2.513 * 10^3$ rad s^{-1};
r = 4 m.

To simplify the calculations, we will estimate that the volume occupied by the gravitational field projected at 45° is given by the sum of six cylinders with the same diameter and height as the Utrons.

$$G = \sqrt{\frac{2U}{y_0 V}} = \sqrt{\frac{2 * 1.2125 * 10^{11}}{1.19230 * 10^9 * 9.817 * 10^{-2}}} = 45.517 \ N \ kg^{-1} = 45.517 \ ms^{-2} .$$

With:

G = Gravitational field [N kg^{-1}] = acceleration [m s^{-2}];
$U = 1.2125 * 10^{11}$ J;
$y_0 = 1.19230 * 10^9$ [$kg^2 \ N^{-1} \ m^{-2}$];
$V = 6 * \pi r^2 * h = 6 * \pi (0.25)^2 * 0.5 = 9.817 * 10^{-2} \ m^3.$

This gravitational field value is found in the area that is 4 meters above the platform and is occupied by both opposing fields. The aforementioned figure is sufficient to carry everything inside the

70

gadget without the effects of inertia and to overcome the strength of the Earth's gravitational field. Within the apparatus, a gravitational field is also created that is centered on the axes of rotation of the two platforms that revolve counterclockwise. The height of the device (estimated here at 3 meters) and the area of the circumference whose radius is the distance from the accumulator to the center of rotation (4 meters in this case) determine the volume occupied by the opposing gravitational fields.

$$G = \sqrt{\frac{2U}{\gamma_0 V}} = \sqrt{\frac{2*1.2125*10^{11}}{1.19230*10^9*1.508*10^2}} = 1.161\, N\, kg^{-1} = 1.161\, m\, s^{-2}$$

With:

G = Gravitational field [N kg^{-1}] = acceleration [m s^{-2}];
$U = 1.2125*10^{11}$ J;
$\gamma_0 = 1.19230*10^9$ [kg^2 N^{-1} m^{-2}];
$V = \pi r^2 * h = \pi(4)^2 * 3 = 1.508 * 10^2$ m^3.

Conclusion

So, there you have it, in conclusion there are significant laws and theories governing the function of the DEC.

The DEC works by spinning gadget parts to generate magnetic fields. Its basic accumulator device is a Faraday's disc application that has been geometrically altered to operate as an energy

71

generator and magnetic propulsion device. It was proved that the created technique can generate energy and propulsion for systems using existing and well-established technologies.

The central energy of the project, that starts the opposed rotation of two platforms, is a set of batteries housed inside the stationary device that function as a generator and produces an intense magnetic field in the center of the craft. Initial calculations with modest dimensions for this field gave $B = 10^6$ T, more than sufficient to recharge the central set of batteries. The peripheral magnetic system, composed of six accumulator devices that pass through the air gap of the external electromagnetic coils, can work as motor or generator. Each accumulator device is a unipolar generator with rotating metallic mercury inside it. Its dimensions, for a craft, can release an excess of energy of 10^{10} J and, depending on rotation, a power of 10^{13} W, with first approximation, although only part of this is necessary to power up the entire device electrical systems.

The metallic capacitor plates distributed in the perimeter of the internal platform accumulate electrotactic charges projected radially and change its atomic structure to become a superconductor. This makes it possible the magnetic buoyancy with the created magnetic field and the terrestrial magnetic field. First calculations shows that it is possible to neutralize the weight of 10^6 kg. The mass of the external platform in rotation may be used to give gyroscopic propulsion, and simple calculations with a 15 m in diameter craft gave the possibility to neutralize the weight of its 25 tons with $4*10^5$ RPM. This is a good result for simple mass rotation.

The possibility to create an artificial gravitational center considering the opposed rotation of two masses showed that the Dave's DEC device used this method as a gravitational shield, to overcome any external gravitational field, turning the craft as a little planet with its own gravity center. It was proven that the technology created can generate energy and propulsion for systems using real, established technology.

Appendix

Python Glowscript Code

Web VPython 3.2

```
scene.background = color.gray(1)

scene.forward = vec(0,-0.2,-1)
```

```
scene.fov = 0.2

scene.range = 3.8

#=====================================================
=========================
#              Inner Platform
#=====================================================
=========================
Bi=sphere(pos=vec(0.26,0.46,0),radius=.01)

Ji=sphere(pos=vec(.65,1.13,0),radius=.01)

Qi=sphere(pos=vec(3.1,3.13,0),radius=.01)

Ri=sphere(pos=vec(1.16,4.25,0),radius=.01)

E1i=sphere(pos=vec(4.25,1.17,0),radius=.01)

Ki=sphere(pos=vec(1.31,0,0),radius=.01)

Wi=sphere(pos=vec(1.11,3.92,0),radius=.01)

Si=sphere(pos=vec(2.84,2.92,0),radius=.01)

Zi=sphere(pos=vec(.97,3.09,0),radius=.01)
```

```
Ti=sphere(pos=vec(2.19,2.39,0),radius=.01)

A1i=sphere(pos=vec(.93,2.83,0),radius=.01)

Vi=sphere(pos=vec(1.99,2.22,0),radius=.01)

H1i=sphere(pos=vec(1.21,1.58,0),radius=.01)

G1i=sphere(pos=vec(1.97,.26,0),radius=.01)

Di=sphere(pos=vec(.5,.78,0),radius=.01)

Ei=sphere(pos=vec(1.24,0,0),radius=.01)

Ii=sphere(pos=vec(.5,-.73,0),radius=.01)


rp1 = extrusion(path=[Bi.pos, Ji.pos],shape=
shapes.circle(radius=.05))

rp2 = extrusion(path=[Ji.pos, Qi.pos],shape=
shapes.circle(radius=.05))

rp3 = extrusion(path=[Qi.pos, Ri.pos],shape=
shapes.circle(radius=.05))

rp4 = extrusion(path=[Ri.pos, Ji.pos],shape=
shapes.circle(radius=.05))
```

```
rp5 = extrusion(path=[Qi.pos, E1i.pos],shape=
shapes.circle(radius=.05))

rp6 = extrusion(path=[E1i.pos, Ki.pos],shape=
shapes.circle(radius=.05))

rp7 = extrusion(path=[Wi.pos, Si.pos],shape=
shapes.circle(radius=.05))

rp8 = extrusion(path=[Zi.pos, Ti.pos],shape=
shapes.circle(radius=.05))

rp9 = extrusion(path=[A1i.pos, Vi.pos],shape=
shapes.circle(radius=.05))

rp10 = extrusion(path=[H1i.pos, G1i.pos],shape=
shapes.circle(radius=.05))

rp11 = extrusion(path=[Di.pos,Ei.pos],shape=
shapes.circle(radius=.05))

rp12 = extrusion(path=[Ei.pos, Ii.pos],shape=
shapes.circle(radius=.05))

rp13 = extrusion(path=[Ii.pos, Di.pos],shape=
shapes.circle(radius=.05))
```

```
#====================================================
============================
#                    Bicone  SCR
#====================================================
============================
c=cone(pos=vector(0,0,0),axis=vector(.35,0,0),radius=.35,

color=color.red,opacity=.8)

c1=cone(pos=vector(0,0,0),axis=vector(-.35,0,0),radius=.35,

color=color.red,opacity=.8)


c8 =sphere(pos=vector(0,0,0),radius=0.2,color=color.gray(1))


c2 = cylinder(pos=vector(0,0,0),axis=vector(.35,0,0), radius=.05)

c3 = cylinder(pos=vector(0,0,0),axis=vector(-.35,0,0), radius=.05)
```

```
c4 = shapes.circle(radius=.08,thickness=.35)  #bearing

c4path = [ vec(.31,0,0), vec(.35,0,0)]

c4e=extrusion(path=c4path, shape=c4,
color=color.red,opacity=1)


c5 = shapes.circle(radius=.08,thickness=.35)  #bearing

c5path = [ vec(-.31,0,0), vec(-.35,0,0)]

c5e=extrusion(path=c5path, shape=c5,
color=color.red,opacity=1)


c6 = box(pos=vector(.33,0,.14),color=color.gray(1),opacity=.8,
    length=.05, height=.3, width=.4)


c7 = box(pos=vector(-.33,0,-.14),color=color.gray(1),opacity=.8,
    length=.05, height=.25, width=.4)
```

```
cb=compound([c,c1,c2,c3,c4e,c5e,c6,c7,c8],pos=vector(3.55,0,0)
)

cb.rotate(angle=pi/4,axis=vec(0,1,0),origin=vector(3.55,0,0))

#=================================================
===========================
#                        Capacitor
#=================================================
===========================


capshape = [ [3.1,3.13], [4.25,1.17], [1.97,.26],
[1.21,1.58],[3.1,3.13]]

cappath = [ vec(0,0,0), vec(0,0,0.1)]

cap=extrusion(path=cappath, shape=capshape,
color=vec(1,.7,.2),opacity=1)



#=================================================
=========================
```

```
#              Inner Platform structure

#=========================================================
=========================


tristruct=compound([rp11,rp12,rp13])

tristruct.rotate(angle=pi/2,axis=vec(1,0,0),origin=vector(0,0,0))


Topring=ring(pos=vector(0,.8,0),axis=vector(0,1,0),radius=0.5,
thickness=0.04)

Topring.rotate(angle=pi/2,axis=vec(1,0,0),origin=vector(0,0,0))

botring=ring(pos=vector(0,-.75,0),axis=vector(0,1,0),radius=0.5,
thickness=0.04)

botring.rotate(angle=pi/2,axis=vec(1,0,0),origin=vector(0,0,0))

centerring=ring(pos=vector(0,0,0),axis=vector(0,1,0),radius=0.5,
thickness=0.04)

centerring.rotate(angle=pi/2,axis=vec(1,0,0),origin=vector(0,0,0))
```

```
centeroutring=ring(pos=vector(0,0,0),axis=vector(0,1,0),radius=
1.25, thickness=0.04)

centeroutring.rotate(angle=pi/2,axis=vec(1,0,0),origin=vector(0,
0,0))


platform1 =
compound([rp1,rp2,rp3,rp4,rp5,rp6,rp7,rp8,rp9,rp10,cap,cb,

tristruct,Topring,botring,centerring,centeroutring]

,origin=vec(0,0,0))

platform12copy = platform1.clone(pos=vector(0,0,0))

platform13copy = platform1.clone(pos=vector(0,0,0))

platform14copy = platform1.clone(pos=vector(0,0,0))

platform15copy = platform1.clone(pos=vector(0,0,0))

platform16copy = platform1.clone(pos=vector(0,0,0))


platform1.rotate(angle=0,axis=vec(0,0,1),origin=vector(0,0,0))
```

```
platform12copy.rotate(angle=1*60*pi/180,axis=vec(0,0,1),origin
=vector(0,0,0))

platform13copy.rotate(angle=2*60*pi/180,axis=vec(0,0,1),origin
=vector(0,0,0))

platform14copy.rotate(angle=3*60*pi/180,axis=vec(0,0,1),origin
=vector(0,0,0))

platform15copy.rotate(angle=4*60*pi/180,axis=vec(0,0,1),origin
=vector(0,0,0))

platform16copy.rotate(angle=5*60*pi/180,axis=vec(0,0,1),origin
=vector(0,0,0))


p1=compound([platform1,platform12copy,platform13copy,platf
orm14copy,

platform15copy,platform16copy,tristruct],origin=vec(0,0,0),

pos=vector(0,0,0),axis=vector(0,1,0))


p1.texture=textures.metal
```

```
#=====================================================
=========================

#              outer Platform

#=====================================================
=========================

rout=ring(pos=vector(0,0,.1),axis=vector(0,0,1),radius=6,
thickness=0.02)

rout.rotate(angle=2*pi,axis=vec(1,0,0),origin=vector(0,0,0))

rout1=ring(pos=vector(0,0,-.16),axis=vector(0,0,1),radius=6,
thickness=0.02)

rout1.rotate(angle=2*pi,axis=vec(1,0,0),origin=vector(0,0,0))


rout2=ring(pos=vector(0,0,1.12),axis=vector(0,0,1),radius=.5,
thickness=0.02)

rout2.rotate(angle=2*pi,axis=vec(1,0,0),origin=vector(0,0,0))
```

```
rout3=ring(pos=vector(0,0,.93),axis=vector(0,0,1),radius=.5,
thickness=0.02)

rout3.rotate(angle=2*pi,axis=vec(1,0,0),origin=vector(0,0,0))


rout4=ring(pos=vector(0,0,-1.06),axis=vector(0,0,1),radius=.5,
thickness=0.02)

rout4.rotate(angle=2*pi,axis=vec(1,0,0),origin=vector(0,0,0))

rout5=ring(pos=vector(0,0,-.88),axis=vector(0,0,1),radius=.5,
thickness=0.02)

rout5.rotate(angle=2*pi,axis=vec(1,0,0),origin=vector(0,0,0))


D=sphere(pos=vec(.5,1.12,0),radius=.01)

e=sphere(pos=vec(.71,1.12,0),radius=.01)

F=sphere(pos=vec(1.15,.67,0),radius=.01)

I=sphere(pos=vec(.86,.95,0),radius=.01)

V=sphere(pos=vec(1.67,.66,0),radius=.01)
```

Z=sphere(pos=vec(2.37,.66,0),radius=.01)

B1=sphere(pos=vec(3.22,.66,0),radius=.01)

D1=sphere(pos=vec(3.92,.66,0),radius=.01)

G=sphere(pos=vec(4.33,.67,0),radius=.01)

F1=sphere(pos=vec(4.94,.47,0),radius=.01)

H1=sphere(pos=vec(5.55,.25,0),radius=.01)

N=sphere(pos=vec(6,.11,0),radius=.01)

H=sphere(pos=vec(6,-.18,0),radius=.01)

I1=sphere(pos=vec(5.5,-.34,0),radius=.01)

K1=sphere(pos=vec(4.93,-.55,0),radius=.01)

M1=sphere(pos=vec(4.39,-.74,0),radius=.01)

O1=sphere(pos=vec(3.76,-.96,0),radius=.01)

J=sphere(pos=vec(3.44,-1.07,0),radius=.01)

Q1=sphere(pos=vec(2.53,-1.08,0),radius=.01)

S1=sphere(pos=vec(1.75,-1.06,0),radius=.01)

```
U1=sphere(pos=vec(1.01,-1.06,0),radius=.01)

W1=sphere(pos=vec(.66,-1.07,0),radius=.01)

K=sphere(pos=vec(.5,-1.07,0),radius=.01)

L=sphere(pos=vec(.5,-.88,0),radius=.01)

V1=sphere(pos=vec(.66,-.88,0),radius=.01)

T1=sphere(pos=vec(1.37,-.88,0),radius=.01)

R1=sphere(pos=vec(2.15,-.88,0),radius=.01)

P1=sphere(pos=vec(2.93,-.88,0),radius=.01)

M=sphere(pos=vec(3.44,-.88,0),radius=.01)

N1=sphere(pos=vec(3.96,-.69,0),radius=.01)

L1=sphere(pos=vec(4.6,-.45,0),radius=.01)

J1=sphere(pos=vec(5.16,-.25,0),radius=.01)

O=sphere(pos=vec(5.83,0,0),radius=.01)

G1=sphere(pos=vec(5.17,0.2,0),radius=.01)

E1=sphere(pos=vec(4.58,.38,0),radius=.01)
```

```
P=sphere(pos=vec(4.33,.46,0),radius=.01)

C1=sphere(pos=vec(3.57,.47,0),radius=.01)

A1=sphere(pos=vec(2.77,.47,0),radius=.01)

W=sphere(pos=vec(2.01,.46,0),radius=.01)

U=sphere(pos=vec(1.33,.47,0),radius=.01)

Q=sphere(pos=vec(1.12,.46,0),radius=.01)

T=sphere(pos=vec(.89,.69,0),radius=.01)

R=sphere(pos=vec(.64,.92,0),radius=.01)

S=sphere(pos=vec(.5,.92,0),radius=.01)

Z1=sphere(pos=vec(1,-.72,0),radius=.01)

C2=sphere(pos=vec(3.18,-.72,0),radius=.01)

B2=sphere(pos=vec(2.55,-.73,0),radius=.01)

A2=sphere(pos=vec(1.76,-.72,0),radius=.01)

E5=sphere(pos=vec(4.58,.38,0),radius=.01)

F5=sphere(pos=vec(4.98,.45,0),radius=.01)
```

```
G5=sphere(pos=vec(5.18,.2,0),radius=.01)

H5=sphere(pos=vec(5.54,.27,0),radius=.01)


F1 = extrusion(path=[D.pos,e.pos],shape=
shapes.circle(radius=.02))

F2 = extrusion(path=[e.pos,F.pos],shape=
shapes.circle(radius=.02))

F3 = extrusion(path=[F.pos,G.pos],shape=
shapes.circle(radius=.02))

F4 = extrusion(path=[G.pos,N.pos], shape=
shapes.circle(radius=.02))

F5 = extrusion(path=[N.pos,H.pos], shape=
shapes.circle(radius=.02))

F6 = extrusion(path=[H.pos,J.pos], shape=
shapes.circle(radius=.02))

F7 = extrusion(path=[J.pos,K.pos], shape=
shapes.circle(radius=.02))
```

```
F8 = extrusion(path=[K.pos,L.pos], shape=
shapes.circle(radius=.02))

F9 = extrusion(path=[L.pos,M.pos], shape=
shapes.circle(radius=.02))

F10 = extrusion(path=[M.pos,O.pos], shape=
shapes.circle(radius=.02))

F11 = extrusion(path=[O.pos,P.pos], shape=
shapes.circle(radius=.02))

F12 = extrusion(path=[P.pos,Q.pos],shape=
shapes.circle(radius=.02))

F13 = extrusion(path=[Q.pos,R.pos],shape=
shapes.circle(radius=.02))

F14 = extrusion(path=[R.pos,S.pos], shape=
shapes.circle(radius=.02))

F15 = extrusion(path=[S.pos,D.pos],shape=
shapes.circle(radius=.02))

F16 = extrusion(path=[V1.pos,Z1.pos],shape=
shapes.circle(radius=.02))
```

```
F17 = extrusion(path=[Z1.pos,N1.pos],shape=
shapes.circle(radius=.02))

F18 = extrusion(path=[R.pos,I.pos],shape=
shapes.circle(radius=.02))

F19 = extrusion(path=[R.pos,e.pos],shape=
shapes.circle(radius=.02))

F20 = extrusion(path=[I.pos,T.pos],shape=
shapes.circle(radius=.02))

F21 = extrusion(path=[T.pos,F.pos],shape=
shapes.circle(radius=.02))

F22 = extrusion(path=[F.pos,U.pos],shape=
shapes.circle(radius=.02))

F23 = extrusion(path=[U.pos,V.pos],shape=
shapes.circle(radius=.02))

F24 = extrusion(path=[V.pos,W.pos],shape=
shapes.circle(radius=.02))

F25 = extrusion(path=[W.pos,Z.pos],shape=
shapes.circle(radius=.02))
```

```
F26 = extrusion(path=[Z.pos,A1.pos],shape=
shapes.circle(radius=.02))

F27 = extrusion(path=[A1.pos,B1.pos],shape=
shapes.circle(radius=.02))

F28 = extrusion(path=[B1.pos,C1.pos],shape=
shapes.circle(radius=.02))

F29 = extrusion(path=[C1.pos,D1.pos],shape=
shapes.circle(radius=.02))

F30 = extrusion(path=[D1.pos,P.pos],shape=
shapes.circle(radius=.02))

F31 = extrusion(path=[P.pos,G.pos],shape=
shapes.circle(radius=.02))

F32 = extrusion(path=[G.pos,E1.pos],shape=
shapes.circle(radius=.02))


#F34 = extrusion(path=[E5.pos,F5.pos],shape=
shapes.circle(radius=.03))
```

```
#F35 = extrusion(path=[F5.pos,G5.pos],shape=
shapes.circle(radius=.03))

#F63 = extrusion(path=[G5.pos,H5.pos],shape=
shapes.circle(radius=.03))

#F64 = extrusion(path=[H5.pos,O.pos],shape=
shapes.circle(radius=.03))


F36 = extrusion(path=[H1.pos,O.pos],shape=
shapes.circle(radius=.02))

F37 = extrusion(path=[O.pos,N.pos],shape=
shapes.circle(radius=.02))

F38 = extrusion(path=[O.pos,I1.pos],shape=
shapes.circle(radius=.02))

F39 = extrusion(path=[I1.pos,J1.pos],shape=
shapes.circle(radius=.02))

F40 = extrusion(path=[H.pos,O.pos],shape=
shapes.circle(radius=.02))
```

```
F41 = extrusion(path=[J1.pos,K1.pos],shape=
shapes.circle(radius=.02))

F42 = extrusion(path=[K1.pos,L1.pos],shape=
shapes.circle(radius=.02))

F43 = extrusion(path=[L1.pos,M1.pos],shape=
shapes.circle(radius=.02))

F44 = extrusion(path=[M1.pos,N1.pos],shape=
shapes.circle(radius=.02))

F45 = extrusion(path=[N1.pos,O1.pos],shape=
shapes.circle(radius=.02))

F46 = extrusion(path=[O.pos,M.pos],shape=
shapes.circle(radius=.02))

F47 = extrusion(path=[M.pos,J.pos],shape=
shapes.circle(radius=.02))

F48 = extrusion(path=[J.pos,P1.pos],shape=
shapes.circle(radius=.02))

F49 = extrusion(path=[P1.pos,Q1.pos],shape=
shapes.circle(radius=.02))
```

```
F50 = extrusion(path=[Q1.pos,R1.pos],shape=
shapes.circle(radius=.02))

F51 = extrusion(path=[R1.pos,S1.pos],shape=
shapes.circle(radius=.02))

F52 = extrusion(path=[S1.pos,T1.pos],shape=
shapes.circle(radius=.02))

F53 = extrusion(path=[T1.pos,U1.pos],shape=
shapes.circle(radius=.02))

F54 = extrusion(path=[V1.pos,W1.pos],shape=
shapes.circle(radius=.02))

F55 = extrusion(path=[U1.pos,V1.pos],shape=
shapes.circle(radius=.02))

F56 = extrusion(path=[Z1.pos,T1.pos],shape=
shapes.circle(radius=.02))

F57 = extrusion(path=[T1.pos,A2.pos],shape=
shapes.circle(radius=.02))

F58 = extrusion(path=[A2.pos,R1.pos],shape=
shapes.circle(radius=.02))
```

```
F59 = extrusion(path=[R1.pos,B2.pos],shape=
shapes.circle(radius=.02))

F60 = extrusion(path=[B2.pos,P1.pos],shape=
shapes.circle(radius=.02))

F61 = extrusion(path=[P1.pos,C2.pos],shape=
shapes.circle(radius=.02))

F62 = extrusion(path=[C2.pos,M.pos],shape=
shapes.circle(radius=.02))




#=====================================================
==========================
#               transformer
#=====================================================
==========================
trans=extrusion(path=[vec(0,0,0), vec(0,0,-.2)],

   shape= shapes.points(pos=[ [3.41777,0.37867],
[3.51641,0.64375],

   [4.50891,0.64375], [4.83564,0.47731],[4.8418,-0.46588],
```

[4.50275,-0.57068],[3.55339,-0.58301],[3.4301,-0.33643]

,[3.59655,-0.33643],[3.68902,-0.45972],[4.61988,-0.33643],

[4.62604,0.391],[3.65203,0.47731],[3.59655,0.37251]]]))


```
circ = shapes.circle(radius=.025)

rectpath = paths.rectangle(width=.25, height=.25)

coil=extrusion(path=rectpath, shape=circ,pos=vector(4.72,.2,-.1),
color=color.red)


coil1 = coil.clone(pos=vector(4.72,.15,-.1))

coil2 = coil.clone(pos=vector(4.72,.1,-.1))

coil3 = coil.clone(pos=vector(4.72,.05,-.1))

coil4 = coil.clone(pos=vector(4.72,0,-.1))

coil5 = coil.clone(pos=vector(4.72,-.05,-.1))

coil6 = coil.clone(pos=vector(4.72,-.1,-.1))
```

```
coil7 = coil.clone(pos=vector(4.72,-.15,-.1))

coil8 = coil.clone(pos=vector(4.72,-.2,-.1))


cabc =
compound([coil,coil1,coil2,coil3,coil4,coil5,coil6,coil7,coil8])

cabc.rotate(angle=pi/2,axis=vec(0,1,0),origin=vector(0,0,0))



#=====================================================
=========================
#              outer Platform structure
#=====================================================
=========================


platform2 =
compound([F1,F2,F3,F4,F5,F6,F7,F8,F9,F10,F11,F12,F13,F14,F15,
F16,
```

```
F17,F18,F19,F20,F21,F22,F23,F24,F25,F26,F27,F28,F29,F30,F31,
F32,

F36,F37,F38,F39,F40,F41,F42,F43,F44,F45,F46,F47,F48,F49,F50,
F51,F52,F53,F54,

F55,F56,F57,F58,F59,F60,F61,F62,trans,cabc],origin=vec(0,0,0),

pos=vector(0,0,0),axis=vector(0,1,0))


platform2.rotate(angle=pi/2,axis=vec(0,1,0),origin=vector(0,0,0))


platform21copy = platform2.clone(pos=vector(0,0,0))

platform22copy = platform2.clone(pos=vector(0,0,0))

platform2.rotate(angle=0,    axis=vec(0,0,1),
origin=vector(0,0,0))

platform21copy.rotate(angle=30*pi/180,    axis=vec(0,0,1),
origin=vector(0,0,0))

platform22copy.rotate(angle=60*pi/180,    axis=vec(0,0,1),
origin=vector(0,0,0))
```

```
platform31copy = platform2.clone(pos=vector(0,0,0))

platform32copy = platform2.clone(pos=vector(0,0,0))

platform33copy = platform2.clone(pos=vector(0,0,0))

platform31copy.rotate(angle=pi/2,    axis=vec(0,0,1),
origin=vector(0,0,0))

platform32copy.rotate(angle=3*30*pi/180,    axis=vec(0,0,1),
origin=vector(0,0,0))

platform33copy.rotate(angle=4*30*pi/180,    axis=vec(0,0,1),
origin=vector(0,0,0))


platform41copy = platform2.clone(pos=vector(0,0,0))

platform42copy = platform2.clone(pos=vector(0,0,0))

platform43copy = platform2.clone(pos=vector(0,0,0))

platform41copy.rotate(angle=5*30*pi/180,    axis=vec(0,0,1),
origin=vector(0,0,0))
```

```
platform42copy.rotate(angle=6*30*pi/180,    axis=vec(0,0,1),
origin=vector(0,0,0))

platform43copy.rotate(angle=7*30*pi/180,    axis=vec(0,0,1),
origin=vector(0,0,0))


platform51copy = platform2.clone(pos=vector(0,0,0))

platform52copy = platform2.clone(pos=vector(0,0,0))

platform53copy = platform2.clone(pos=vector(0,0,0))

platform54copy = platform2.clone(pos=vector(0,0,0))

platform51copy.rotate(angle=8*30*pi/180,    axis=vec(0,0,1),
origin=vector(0,0,0))

platform52copy.rotate(angle=9*30*pi/180,    axis=vec(0,0,1),
origin=vector(0,0,0))

platform53copy.rotate(angle=10*30*pi/180,    axis=vec(0,0,1),
origin=vector(0,0,0))

platform54copy.rotate(angle=11*30*pi/180,    axis=vec(0,0,1),
origin=vector(0,0,0))
```

```
p2=compound([platform2,platform21copy,platform22copy,platf
orm31copy,platform32copy,

platform33copy,platform41copy,platform42copy,platform43cop
y,platform51copy,

platform52copy,platform53copy,platform54copy,rout],origin=ve
c(0,0,0), pos=vector(0,0,0)

,axis=vector(0,1,0))

p2.texture=textures.metal




#=====================================================
============================
#              Center structure (stationary)
#=====================================================
============================
```

```
cr = extrusion(path=[vec(0,0,-1.7),vec(0,0,1.3)],

    shape= shapes.circle(radius=.25,thickness=.1))



#=====================================================
============================
#           Searl Bearings
#=====================================================
============================
#data points
ic=sphere(pos=vec(0,1.31,0),radius=.01)
ec=sphere(pos=vec(0,1.05,0),radius=.01)
dc=sphere(pos=vec(0,.6,0),radius=.01)
cc=sphere(pos=vec(0,0,0),radius=.01)
fc=sphere(pos=vec(0,-1.52,0),radius=.01)
gc=sphere(pos=vec(0,1,0),radius=.01)
hc=sphere(pos=vec(0,-1.74,0),radius=.01)
```

```
jc=sphere(pos=vec(.25,-1.2,0),radius=.01)

kc=sphere(pos=vec(1.96,-1.2,0),radius=.01)

oc=sphere(pos=vec(1.96,-1.62,0),radius=.01)

nc=sphere(pos=vec(1.42,-1.62,0),radius=.01)

mc=sphere(pos=vec(.91,-1.74,0),radius=.01)

lc=sphere(pos=vec(.25,-1.74,0),radius=.01)


r1=extrusion(path=[vec(0,0,0), vec(0,0,2/12)],
color=color.gray(.2),

shape=[ shapes.circle(radius=3/12),
shapes.circle(radius=3.25/12) ])

r2=extrusion(path=[vec(0,0,0), vec(0,0,2/12)],
color=color.yellow,

shape=[ shapes.circle(radius=3.25/12),
shapes.circle(radius=4/12) ])

r3=extrusion(path=[vec(0,0,0), vec(0,0,2/12)],
color=color.orange,
```

```
shape=[ shapes.circle(radius=4/12),
shapes.circle(radius=4.25/12) ])

r4=extrusion(path=[vec(0,0,0), vec(0,0,2/12)],
color=color.orange,

shape=[ shapes.circle(radius=5.25/12),
shapes.circle(radius=5.5/12) ])


r5=extrusion(path=[vec(4.75/12,0,0),
vec(4.75/12,0,2/12)],shape=[ shapes.circle(radius=.375/12)])

r5.rotate(angle=0,axis=vec(0,0,1),origin=vector(0,0,0))

r6=extrusion(path=[vec(4.75/12,0,0),
vec(4.75/12,0,2/12)],shape=[ shapes.circle(radius=.375/12)])

r6.rotate(angle=15*pi/180,axis=vec(0,0,1),origin=vector(0,0,0))

r7=extrusion(path=[vec(4.75/12,0,0),
vec(4.75/12,0,2/12)],shape=[ shapes.circle(radius=.375/12)])

r7.rotate(angle=30*pi/180,axis=vec(0,0,1),origin=vector(0,0,0))

r8=extrusion(path=[vec(4.75/12,0,0),
vec(4.75/12,0,2/12)],shape=[ shapes.circle(radius=.375/12)])
```

```
r8.rotate(angle=45*pi/180,axis=vec(0,0,1),origin=vector(0,0,0))

r9=extrusion(path=[vec(4.75/12,0,0),
vec(4.75/12,0,2/12)],shape=[ shapes.circle(radius=.375/12)]])

r9.rotate(angle=60*pi/180,axis=vec(0,0,1),origin=vector(0,0,0))

r10=extrusion(path=[vec(4.75/12,0,0),
vec(4.75/12,0,2/12)],shape=[ shapes.circle(radius=.375/12)]])

r10.rotate(angle=75*pi/180,axis=vec(0,0,1),origin=vector(0,0,0))

r11=extrusion(path=[vec(4.75/12,0,0),
vec(4.75/12,0,2/12)],shape=[ shapes.circle(radius=.375/12)]])

r11.rotate(angle=90*pi/180,axis=vec(0,0,1),origin=vector(0,0,0))

r12=extrusion(path=[vec(4.75/12,0,0),
vec(4.75/12,0,2/12)],shape=[ shapes.circle(radius=.375/12)]])

r12.rotate(angle=105*pi/180,axis=vec(0,0,1),origin=vector(0,0,0)
)

r13=extrusion(path=[vec(4.75/12,0,0),
vec(4.75/12,0,2/12)],shape=[ shapes.circle(radius=.375/12)]])

r13.rotate(angle=120*pi/180,axis=vec(0,0,1),origin=vector(0,0,0)
)
```

```
r14=extrusion(path=[vec(4.75/12,0,0),
vec(4.75/12,0,2/12)],shape=[ shapes.circle(radius=.375/12)])

r14.rotate(angle=135*pi/180,axis=vec(0,0,1),origin=vector(0,0,0)
)

r15=extrusion(path=[vec(4.75/12,0,0),
vec(4.75/12,0,2/12)],shape=[ shapes.circle(radius=.375/12)])

r15.rotate(angle=150*pi/180,axis=vec(0,0,1),origin=vector(0,0,0)
)

r16=extrusion(path=[vec(4.75/12,0,0),
vec(4.75/12,0,2/12)],shape=[ shapes.circle(radius=.375/12)])

r16.rotate(angle=165*pi/180,axis=vec(0,0,1),origin=vector(0,0,0)
)

r17=extrusion(path=[vec(4.75/12,0,0),
vec(4.75/12,0,2/12)],shape=[ shapes.circle(radius=.375/12)])

r17.rotate(angle=180*pi/180,axis=vec(0,0,1),origin=vector(0,0,0)
)

r18=extrusion(path=[vec(4.75/12,0,0),
vec(4.75/12,0,2/12)],shape=[ shapes.circle(radius=.375/12)])
```

```
r18.rotate(angle=195*pi/180,axis=vec(0,0,1),origin=vector(0,0,0)
)

r20=extrusion(path=[vec(4.75/12,0,0),
vec(4.75/12,0,2/12)],shape=[ shapes.circle(radius=.375/12)])

r20.rotate(angle=210*pi/180,axis=vec(0,0,1),origin=vector(0,0,0)
)

r21=extrusion(path=[vec(4.75/12,0,0),
vec(4.75/12,0,2/12)],shape=[ shapes.circle(radius=.375/12)])

r21.rotate(angle=225*pi/180,axis=vec(0,0,1),origin=vector(0,0,0)
)

r22=extrusion(path=[vec(4.75/12,0,0),
vec(4.75/12,0,2/12)],shape=[ shapes.circle(radius=.375/12)])

r22.rotate(angle=240*pi/180,axis=vec(0,0,1),origin=vector(0,0,0)
)

r23=extrusion(path=[vec(4.75/12,0,0),
vec(4.75/12,0,2/12)],shape=[ shapes.circle(radius=.375/12)])

r23.rotate(angle=255*pi/180,axis=vec(0,0,1),origin=vector(0,0,0)
)
```

```
r24=extrusion(path=[vec(4.75/12,0,0),
vec(4.75/12,0,2/12)],shape=[ shapes.circle(radius=.375/12)])

r24.rotate(angle=270*pi/180,axis=vec(0,0,1),origin=vector(0,0,0)
)

r25=extrusion(path=[vec(4.75/12,0,0),
vec(4.75/12,0,2/12)],shape=[ shapes.circle(radius=.375/12)])

r25.rotate(angle=285*pi/180,axis=vec(0,0,1),origin=vector(0,0,0)
)

r26=extrusion(path=[vec(4.75/12,0,0),
vec(4.75/12,0,2/12)],shape=[ shapes.circle(radius=.375/12)])

r26.rotate(angle=300*pi/180,axis=vec(0,0,1),origin=vector(0,0,0)
)

r27=extrusion(path=[vec(4.75/12,0,0),
vec(4.75/12,0,2/12)],shape=[ shapes.circle(radius=.375/12)])

r27.rotate(angle=315*pi/180,axis=vec(0,0,1),origin=vector(0,0,0)
)

r28=extrusion(path=[vec(4.75/12,0,0),
vec(4.75/12,0,2/12)],shape=[ shapes.circle(radius=.375/12)])
```

```
r28.rotate(angle=330*pi/180,axis=vec(0,0,1),origin=vector(0,0,0)
)

r29=extrusion(path=[vec(4.75/12,0,0),
vec(4.75/12,0,2/12)],shape=[ shapes.circle(radius=.375/12)])

r29.rotate(angle=345*pi/180,axis=vec(0,0,1),origin=vector(0,0,0)
)

#r30=extrusion(path=[vec(4.75/12,0,0),
vec(4.75/12,0,2/12)],shape=[ shapes.circle(radius=.375/12)])

#r30.rotate(angle=360*pi/180,axis=vec(0,0,1),origin=vector(0,0,
0))


bb =
compound([r1,r2,r3,r4,r5,r6,r7,r8,r9,r10,r11,r12,r13,r14,r15,r16,
r17,r18,

r20,r21,r22,r23,r24,r25,r26,r27,r28,r29],pos=vector(0,0,1))


bb2=bb.clone(pos=vector(0,0,-1))

bb3=bb.clone(pos=vector(0,0,-.6))
```

```
bb4=bb.clone(pos=vector(0,0,.6))

#bearing=compound([bb,bb2,bb3,bb4])

#=====================================================
============================

#            biefel-Brown spheres

#=====================================================
============================

biefeldbrownl = sphere(pos=vector(2.3,0,-1.5), radius=0.4,)

biefeldbrownl.rotate(angle=pi/3,

    axis=vec(0,0,1),

    origin=vector(0,0,0))

biefeldbrownlcopy = biefeldbrownl.clone(pos=vector(2.3,0,-1.5))

biefeldbrownlcopy.rotate(angle=5*pi/3,

    axis=vec(0,0,1),

    origin=vector(0,0,0))

biefeldbrown2copy = biefeldbrownl.clone(pos=vector(2.3,0,-1.5))
```

```
biefeldbrown2copy.rotate(angle=9*pi/3,

        axis=vec(0,0,1),

        origin=vector(0,0,0))



#=====================================================
===========================

#               tri leg structure

#=====================================================
===========================

Ac=sphere(pos=vec(0.250080000000000, -
1.21555000000000,0),radius=.01)

Bc=sphere(pos=vec(1.95557246730073, -
1.22318781751494,0),radius=.01)

Cc=sphere(pos=vec(1.94792995880150, -
1.62059825947455,0),radius=.01)

Dc=sphere(pos=vec(0.250000000000000, -
1.62824076797378,0),radius=.01)
```

```
rp1 = extrusion(path=[Ac.pos, Bc.pos],shape=
shapes.circle(radius=.05))

rp2 = extrusion(path=[Bc.pos, Cc.pos],shape=
shapes.circle(radius=.05))

rp3 = extrusion(path=[Cc.pos, Dc.pos],shape=
shapes.circle(radius=.05))

rp4 = extrusion(path=[Cc.pos, Ac.pos],shape=
shapes.circle(radius=.05))


bt1 = compound([rp1,rp2,rp3,rp4],origin=vec(0,0,0))

bt1.rotate(angle=pi/2,axis=vec(1,0,0),origin=vector(0,0,0))

bt1.rotate(angle=60*pi/180, axis=vec(0,0,1),
origin=vector(0,0,0))


bt12c = bt1.clone(pos=vector(0,0,0))
```

```
bt12c.rotate(angle=120*pi/180,
axis=vec(0,0,1),origin=vector(0,0,0))


bt13c = bt1.clone(pos=vector(0,0,0))

bt13c.rotate(angle=240*pi/180,
axis=vec(0,0,1),origin=vector(0,0,0))


ringB=ring(pos=vector(0,0,-1.2),axis=vector(0,0,1), radius=2,
thickness=0.051)


centerpiece=compound([cr,bt1,bt12c,bt13c,biefeldbrownl,

biefeldbrownlcopy,biefeldbrown2copy,ringB])


#=====================================================
============================
#              Lower Cover Frame structure
```

```
#=====================================================
===========================
```

Ab=sphere(pos=vec(0.292720000000000,
1.25589000000000,0),radius=.01)

Bb=sphere(pos=vec(0.921540000000001,
1.26514000000000,0),radius=.01)

Cb=sphere(pos=vec(1.37466539627476,
0.812012195828232,0),radius=.01)

Db=sphere(pos=vec(4.12115000000000,
0.802760000000001,0),radius=.01)

Eb=sphere(pos=vec(6.10935226171360,
0.146196855375896,0),radius=.01)

Fb=sphere(pos=vec(6.09085739114548, -
0.186710814850272,0),radius=.01)

Gb=sphere(pos=vec(2.09596534843146, -
1.62931071916367,0),radius=.01)

```
Hb=sphere(pos=vec(0.301962903323775, -
1.64780558973179,0),radius=.01)


rp1 = extrusion(path=[Ab.pos, Bb.pos],shape=
shapes.circle(radius=.02))

rp2 = extrusion(path=[Bb.pos, Cb.pos],shape=
shapes.circle(radius=.02))

rp3 = extrusion(path=[Cb.pos, Db.pos],shape=
shapes.circle(radius=.02))

rp4 = extrusion(path=[Db.pos, Eb.pos],shape=
shapes.circle(radius=.02))

rp5 = extrusion(path=[Eb.pos, Fb.pos],shape=
shapes.circle(radius=.02))

rp6 = extrusion(path=[Fb.pos, Gb.pos],shape=
shapes.circle(radius=.02))

rp7 = extrusion(path=[Gb.pos, Hb.pos],shape=
shapes.circle(radius=.02))
```

```
base1 = compound([rp1,rp2,rp3,rp4,rp5,rp6,rp7],

origin=vec(0,0,0))

base1.rotate(angle=pi/2,axis=vec(1,0,0),origin=vector(0,0,0))


base12copy = base1.clone(pos=vector(0,0,0))

base13copy =  base1.clone(pos=vector(0,0,0))

base14copy =  base1.clone(pos=vector(0,0,0))

base15copy =  base1.clone(pos=vector(0,0,0))

base16copy =  base1.clone(pos=vector(0,0,0))


base1.rotate(angle=0,axis=vec(0,0,1),origin=vector(0,0,0))

base12copy.rotate(angle=1*60*pi/180,axis=vec(0,0,1),origin=ve
ctor(0,0,0))

base13copy.rotate(angle=2*60*pi/180,axis=vec(0,0,1),origin=ve
ctor(0,0,0))
```

```
base14copy.rotate(angle=3*60*pi/180,axis=vec(0,0,1),origin=ve
ctor(0,0,0))

base15copy.rotate(angle=4*60*pi/180,axis=vec(0,0,1),origin=ve
ctor(0,0,0))

base16copy.rotate(angle=5*60*pi/180,axis=vec(0,0,1),origin=ve
ctor(0,0,0))


r1=ring(pos=vector(0,0,.8), axis=vector(0,0,1),radius=4.15,
thickness=0.03)

r1.rotate(angle=pi,axis=vec(0,0,1),origin=vector(0,0,0))


r2=ring(pos=vector(0,0,.15), axis=vector(0,0,1),radius=6.1,
thickness=0.03)

r2.rotate(angle=pi,axis=vec(0,0,1),origin=vector(0,0,0))


r3=ring(pos=vector(0,0,-.2), axis=vector(0,0,1),radius=6.1,
thickness=0.03)
```

```
r3.rotate(angle=pi,axis=vec(0,0,1),origin=vector(0,0,0))


pb1=compound([base1,base12copy,base13copy,base14copy,bas
e15copy,base16copy,r1,

r2,r3],origin=vec(0,0,0),pos=vector(0,0,0),axis=vector(0,1,0))


pb1.texture=textures.metal


#====================================================
============================
#              Upper Cover Frame structure
#====================================================
============================


lb=sphere(pos=vec(4.12115000000000,
0.802760000000001,0),radius=.01)
```

```
Jb=sphere(pos=vec(5.48915622123752,
1.27953956956276,0),radius=.01)

Kb=sphere(pos=vec(4.44254000000001,
2.07456000000000,0),radius=.01)

Lb=sphere(pos=vec(3.74815353576465,
2.54755308615001,0),radius=.01)

Mb=sphere(pos=vec(2.97609732086995,
2.95436552627399,0),radius=.01)

Nb=sphere(pos=vec(2.00043157135653,
3.32004643169670,0),radius=.01)

Ob=sphere(pos=vec(1.02742750151019,
3.54168413751754,0),radius=.01)

Pb=sphere(pos=vec(0.00000000000000,
3.63442181465336,0),radius=.01)


rp8 = extrusion(path=[lb.pos, Jb.pos],shape=
shapes.circle(radius=.02))
```

```
rp9 = extrusion(path=[Jb.pos, Kb.pos],shape=
shapes.circle(radius=.02))

rp10 = extrusion(path=[Kb.pos, Lb.pos],shape=
shapes.circle(radius=.02))

rp11 = extrusion(path=[Lb.pos, Mb.pos],shape=
shapes.circle(radius=.02))

rp12 = extrusion(path=[Mb.pos, Nb.pos],shape=
shapes.circle(radius=.02))

rp13 = extrusion(path=[Nb.pos, Ob.pos],shape=
shapes.circle(radius=.02))

rp14 = extrusion(path=[Ob.pos, Pb.pos],shape=
shapes.circle(radius=.02))


base1 = compound([rp8,rp9,rp10,rp11,rp12,rp13,rp14],

origin=vec(0,0,0))

base1.rotate(angle=pi/2,axis=vec(1,0,0),origin=vector(0,0,0))
```

```
base12copy = base1.clone(pos=vector(0,0,0))

base13copy =  base1.clone(pos=vector(0,0,0))

base14copy =  base1.clone(pos=vector(0,0,0))

base15copy =  base1.clone(pos=vector(0,0,0))

base16copy =  base1.clone(pos=vector(0,0,0))


base1.rotate(angle=0,axis=vec(0,0,1),origin=vector(0,0,0))

base12copy.rotate(angle=1*60*pi/180,axis=vec(0,0,1),origin=ve
ctor(0,0,0))

base13copy.rotate(angle=2*60*pi/180,axis=vec(0,0,1),origin=ve
ctor(0,0,0))

base14copy.rotate(angle=3*60*pi/180,axis=vec(0,0,1),origin=ve
ctor(0,0,0))

base15copy.rotate(angle=4*60*pi/180,axis=vec(0,0,1),origin=ve
ctor(0,0,0))

base16copy.rotate(angle=5*60*pi/180,axis=vec(0,0,1),origin=ve
ctor(0,0,0))
```

```
r3=ring(pos=vector(0,0,3), axis=vector(0,0,1),radius=3,
thickness=0.03)

r3.rotate(angle=pi,axis=vec(0,0,1),origin=vector(0,0,0))


r4=ring(pos=vector(0,0,1.3), axis=vector(0,0,1),radius=5.5,
thickness=0.03)

r4.rotate(angle=pi,axis=vec(0,0,1),origin=vector(0,0,0))


ucf=compound([base1,base12copy,base13copy,base14copy,bas
e15copy,base16copy,

r3,r4],

origin=vec(0,0,0),pos=vector(0,0,0),axis=vector(0,1,0),opacity=(
2))


ucf.texture=textures.metal
```

```
#===========================covering=================
==========================

fr = [ [0.0, 1.45], [0.0, 1.3478], [0.775, 1.345], [0.869,
1.448],[0.869, 1.448]

,[1.3, 0.77],[1.402, 0.876],[4.505, 0.889],[4.498, 0.769],[6.299,
0.196],

[6.159, -0.302],[6.173, 0.146],[6.081, -0.204],[2.879, -
1.447],[2.865, -1.313],

[2.289, -1.082],[2.219, -1.194],[0.0, -1.181],[0.0, -1.092],[0.0,
1.45]]


circpath = paths.circle(radius=.1)

cover=extrusion(path=circpath, shape=fr,
color=color.yellow,opacity=.3,scale=1)

cover.rotate(angle=pi/2, axis=vec(1,0,0), origin=vector(0,0,0))

##color=vec(1,.7,.2)
```

```
#=====================================================
==========================

top=[[0,3.63],[1.03,3.54],[2,3.32],[2.98,2.95],[2.87,2.88],[1.97,3.
22],

[1.01,3.43],[0,3.51],[0,3.63]]


circpath = paths.circle(radius=.1)

cover=extrusion(path=circpath, shape=top, color=color.gray(.1),

opacity=.1,scale=1)

cover.rotate(angle=pi/2, axis=vec(1,0,0), origin=vector(0,0,0))

##color=color.gray(.1)

#=====================================================
==========================

front=[[4.44,2.07],[4.36,1.97],[5.25,1.3],[4.12,.91],[4.12,.8],[5.49,
1.28],

[4.44,2.07]]
```

```
circpath = paths.circle(radius=.1)

cover=extrusion(path=circpath, shape=front,
color=color.yellow,opacity=.3

,scale=1)

cover.rotate(angle=pi/2, axis=vec(1,0,0), origin=vector(0,0,0))

##color=color.gray(.1)

#=====================================================
==========================

top=[[0,3.63],[1.03,3.54],[2,3.32],[2.98,2.95],[2.87,2.88],[1.97,3.
22],

[1.01,3.43],[0,3.51],[0,3.63]]


circpath = paths.circle(radius=.1)

cover=extrusion(path=circpath, shape=top,
color=color.yellow,opacity=.3

,scale=1)

cover.rotate(angle=pi/2, axis=vec(1,0,0), origin=vector(0,0,0))
```

```
#=======================================================
========================

#=======================================================
========================

win=[[2.98,2.95],[3.75,2.55],[4.44,2.07],[4.36,1.97],[3.69,2.45],[
2.87,2.88],

[2.98,2.95]]


circpath = paths.circle(radius=.1)

cover=extrusion(path=circpath, shape=win,
color=color.cyan,opacity=.1,scale=1)

cover.rotate(angle=pi/2, axis=vec(1,0,0), origin=vector(0,0,0))

#=======================================================
===========================


#=======================================================
===========================
```

```
run = True

def runner(r):
    global run
    run = r.checked

checkbox(bind=runner, text='Run', checked=True)
scene.waitfor('textures')

t = 0
dt = 0.01
dtheta = 0.01
while True:
    rate(10000)
    if run:
```

```
p1.rotate(angle=-dtheta, axis=vec(0,0,1))

p2.rotate(angle=dtheta, axis=vec(0,0,1))

bb.rotate(angle=-dtheta, axis=vec(0,0,1))

bb2.rotate(angle=-dtheta, axis=vec(0,0,1))

bb3.rotate(angle=-dtheta, axis=vec(0,0,1))

bb4.rotate(angle=dtheta, axis=vec(0,0,1))

t += dt
```

www.ingramcontent.com/pod-product-compliance
Lightning Source LLC
Chambersburg PA
CBHW062322290526
45794CB00005B/1859